NICHOLAS MIRZOEFF

How to See the World

A PELICAN INTRODUCTION

PELICAN
an imprint of
PENGUIN BOOKS

PELICAN BOOKS

UK | USA | Canada | Ireland | Australia
India | New Zealand | South Africa

Penguin Books is part of the Penguin Random House
group of companies whose addresses can be found at
global.penguinrandomhouse.com.

Penguin
Random House
UK

First published 2015

011

Text copyright © Nicholas Mirzoeff, 2015
The moral right of the author has been asserted

Book design by Matthew Young
Set in 10/14.664pt Freight Text Pro
Printed and bound in Great Britain by Clays Ltd, Elcograf S.p.A.

A CIP catalogue record for this book is available from
the British Library

ISBN: 978-0-141-97740-9

MIX
Paper | Supporting
responsible forestry
FSC® C018179

Penguin Random House is committed to a sustainable future for our business, our readers and our
planet. This book is made from Forest Stewardship
Council® certified paper.

www.greenpenguin.co.uk

Contents

ACKNOWLEDGEMENTS

As ever, I owe everything to Kathleen, and to Hannah, who showed me how teenagers are remaking the world through social media. My thanks are due to Laura Stickney for commissioning this book and seeing it through various stages of conversion from an academic book to one that might see the light of day as a trade publication. Many thanks to Monica Schmoller for empathetic and intelligent copy-editing. All inadequacies that remain are mine, of course. The ideas in the book result from my interactions with people at New York University and Middlesex University, where I teach and work – as well as many other places which I have had the opportunity to visit – and I am grateful to everyone at both colleges. I want to acknowledge from the outset my intellectual debt to John Berger, to feminist approaches to visual culture and to British cultural studies. In that light, I'd like to dedicate this book to the memory of cultural theorist Stuart Hall, a mentor and inspiration to so many of us: Rest In Power.

Nicholas Mirzoeff

How to See the World

Figure 1 — NASA, *Blue Marble*

In 1972, the astronaut Jack Schmitt took a picture of Earth from the *Apollo 17* spacecraft, which is now believed to be the most reproduced photograph ever. Because it showed the spherical globe dominated by blue oceans with intervening green landmasses and swirling clouds, the image came to be known as *Blue Marble*.

The photograph powerfully depicted the planet as a whole, and from space: no human activity or presence was visible. It appeared on almost every newspaper front page around the world.

In the photograph, Earth is viewed very close to the edge

of the frame. It dominates the picture and overwhelms our senses. Since the spacecraft had the sun behind it, the photograph was unique in showing the planet fully illuminated. The Earth seems at once immense and knowable. Taught to recognize the outline of the continents, viewers could now see how these apparently abstract shapes were a lived and living whole. The photograph mixed the known and the new in a visual format that made it comprehensible and beautiful.

At the time it was published, many people believed that seeing *Blue Marble* changed their lives. The poet Archibald MacLeish recalled that for the first time people saw the Earth as a whole, 'whole and round and beautiful and small'. Some found spiritual and environmental lessons in viewing the planet as if from the place of a god. Writer Robert Poole called *Blue Marble* 'a photographic manifesto for global justice' (Wuebbles 2012). It inspired utopian thoughts of a world government, perhaps even a single global language, epitomized by its use on the front cover of *The Whole Earth Catalog*, the classic book of the counterculture. Above all, it seemed to show that the world was a single, unified place. As Apollo astronaut Russell ('Rusty') Schweickart put it, the image conveys

> the thing is a whole, the Earth is a whole, and it's so beautiful. You wish you could take a person in each hand, one from each side in the various conflicts, and say, 'Look. Look at it from this perspective. Look at that. What's important?'

No human has seen that perspective in person since the photograph was taken, yet most of us feel we know how the Earth looks because of *Blue Marble*.

That unified world, visible from one spot, often seems out of reach now. In the forty years since *Blue Marble*, the world has changed dramatically in four key registers. Today, the world is young, urban, wired and hot. Each of these indicators has passed a crucial threshold since 2008. In that year, more people lived in cities than the countryside for the first time in history. Consider the emerging world power Brazil. In 1960, only a third of its people lived in cities. By 1972, when *Blue Marble* was taken, the urban population had already passed 50 percent. Today, 85 percent of Brazilians live in cities, no less than 166 million people.

Most of them are young, which is the next indicator. By 2011, more than half the world's population was under thirty; 62 percent of Brazilians are twenty-nine or younger. More than half of the 1.2 billion Indians are under twenty-five, and a similar young majority exists in China. Two-thirds of South Africa's population is under thirty-five. According to the Kaiser Family Foundation, 52 percent of the 18 million people in Niger are under fifteen and in most of sub-Saharan Africa, over 40 percent of the population is under fifteen. The populations of North America, Western Europe and Japan may be ageing, but the global pattern is clear.

The third threshold is connectivity. In 2012, more than a third of the world's population had access to the Internet, up 566 percent since 2000. It's not just Europe and America that are connected: 45 percent of those with Internet access are in Asia. Nonetheless, the major regions that lack connection are sub-Saharan Africa (other than South Africa) and the Indian sub-continent, creating a digital divide on a global level. By the end of 2014, an estimated 3 billion people were online. By the end of the decade, Google envisages 5 billion people on the Internet. This is not just another form of mass media. It is the first universal medium.

One of the most notable uses of the global network is to create, send and view images of all kinds, from photographs to video, comics, art and animation. The numbers are astonishing: one hundred hours of YouTube video are uploaded every minute. Six billion hours of video are watched every month on the site, one hour for every person on earth. The 18–34 age group watches more YouTube than cable television. (And remember that YouTube was only created in 2005.) Every two minutes, Americans alone take more photographs than were made in the entire nineteenth century. As early as 1930, an estimated one billion photographs were being taken every year worldwide. Fifty years later, it was about 25 billion a year, still taken on film. By 2012, we were taking 380 billion photographs a year, nearly all digital. One trillion photographs were taken in 2014. There were some 3.5 trillion photographs in existence in 2011, so the global photography archive increased by some 25 percent or so in 2014. In that same year, 2011, there were one trillion visits to YouTube. Like it or not, the emerging global society is visual. All these photographs and videos are our way of trying to see the world. We feel compelled to make images of it and share them with others as a key part of our effort to understand the changing world around us and our place within it.

The planet itself is changing before our eyes. In 2013, carbon dioxide passed the signature threshold of 400 parts-per-million in the atmosphere for the first time since the Pliocene era about three to five million years ago. Although we cannot see the gas, it has set in motion catastrophic change. With more carbon dioxide, warm air holds more water vapour. As the ice-caps melt, there is more water in the ocean. As the oceans warm, there is more energy for a storm system to draw on, producing storm after 'unprecedented' storm. If a hurricane or earthquake creates what scientists call a 'high sea-level event', like a storm surge or tsunami, the effects are dramatically multiplied. Record-setting floods have followed around the world from Bangkok to London and New York, even as other areas – from Australia to Brazil, California and equatorial Africa – suffer unprecedented drought. The world today is physically different from the one we see in *Blue Marble*, and it is changing fast.

For all the new visual material, it is often hard to be sure what we are seeing when we look at today's world. None of these changes are settled or stable. It seems as if we live in a time of permanent revolution. If we put together these factors of growing, networked cities with a majority youthful population, and a changing climate, what we get is a formula for change. Sure enough, people worldwide are actively trying to change the systems that represent us in all senses, from artistic to visual and political. This book seeks to understand the changing world to help them and all those trying to make sense of what they see.

To get an impression of the distance we have come since *Blue Marble*, consider two photographs from space taken in 2012. In December 2012 the Japanese astronaut Aki Hoshide took his own picture in space. Ignoring the spectacle of Earth, space and moon, Hoshide turned the camera on himself, creating the ultimate 'selfie', or self-taken self-portrait.

Figure 2 — Hoshide, 'Untitled', selfie

Ironically, any trace of his appearance or personality disappears in this image as his reflective visor shows us only what he is looking at – the International Space Station and below it, the Earth. Where *Blue Marble* showed us the planet, Hoshide wants us to see just him. It is nonetheless an undeniably compelling image. By echoing the daily practice of the selfie, the camera and the picture make space real and imaginable to us in an even more direct way than *Blue Marble*, but with none of the social impact of the earlier image. The astronaut is invisible and unknowable in his own self-portrait. There is, it seems, more to seeing than being in the place to see.

In that same year, 2012, NASA created a new version of *Blue Marble*. The new photograph was actually a composite assembled from a series of digital images produced by a satellite. From the satellite's orbit, approximately 930 kilometres (580 miles) above the surface, the full view of the

Figure 3 — NASA, *Blue Marble 2012*

planet is not in fact visible. You have to go over 11,000 kilometres (7,000 miles) away before the entire globe can be seen. The resulting colour-corrected 'photograph', adjusted to show the United States rather than Africa, is now one of the most accessed images on the digital photo archive Flickr, with over five million downloads.

We can 'recognize' the Earth from *Blue Marble*, but only the three-man crew of Apollo 17 have ever actually seen this view, with the earth fully illuminated, and no one has seen it since 1972. The 2012 *Blue Marble* is made to seem as if it was taken from one place in space but it was not. It is accurate in each detail, but it is false in that it gives the illusion of having being taken from a specific place at one moment in time. Such 'tiled rendering' is a standard means of constructing digital imagery. It is a good metaphor for how the world is visualized today. We assemble a world from pieces, assuming that what we see is both coherent and equivalent to reality. Until we discover it is not.

A striking demonstration of how what seems to be a solid whole is actually a composite of assembled pieces came with the 2008 financial crash. What mainstream economists and governments alike had asserted to be the perfectly calculated, global financial market collapsed without warning. It turned out that the system was so finely leveraged that a relatively small number of people, who were unable to keep up with their mortgages, set in motion a rolling catastrophe. The very connectedness of the global financial market made it impossible to contain what would once have been a local misfortune. The crisis shows that it is one world now, like it or not.

At the same time, 'one world' does not mean it is equally available to all. Moving country for personal or political reasons is often very difficult, and partly depends on your passport. A British-passport holder can visit 167 countries without a visa. An Iranian passport, however, gets you into only 46 countries. Money, on the other hand, can move wherever it wants at the click of a keyboard. Prior to 1979, it was illegal for Chinese citizens to even possess foreign currency. Today China dominates global trade. There is globalization in theory, which is smooth and easy. And there is the uneven, difficult and time-consuming experience of globalization in practice. The ads and the politicians tell us there is a single global system now, at least for financial affairs. Our daily lives tell us otherwise.

Visual culture

This book is designed to help you see the much-changed and changing world. It is a guide to the visual culture we live in. Like history, visual culture is both the name of the academic field and that of its object of study. Visual culture involves the things that we see, the mental model we all have of how to see, and what we can do as a result. That is why we call it visual culture: a culture of the visual. A visual culture is not simply the total amount of what has been made to be seen, such as paintings or films. A visual culture is the relation between what is visible and the names that we give to what is seen. It also involves what is invisible or kept out of sight. In short, we don't simply see what there is to see and call it a visual culture. Rather, we assemble a world-view that is consistent with what we know and have already experienced. There are institutions that try to shape that view, which the French historian Jacques Rancière calls 'the police version of history', meaning that we are told to 'move on, there's nothing to see here' (2001). Only of course there is something to see, we just usually choose to let the authorities deal with the situation. If it is a traffic accident, that may be appropriate. If it is a question of how we see history as a whole, then surely we should be looking.

The concept of visual culture as a specific area of study first began to circulate at a previous moment of vital change in the way we see the world. Around 1990, the end of the Cold War that had divided the globe into two zones, more or less invisible to each other, coincided with the rise of what was called 'postmodernism'. The postmodern changed modern skyscrapers from austere rectangular blocks into the playful towers, with kitschy and pastiche features, that now dominate skylines worldwide. Cities looked very different. A new identity politics formed around questions of gender, sexuality and race, leading people to see themselves differently. This politics was less confident in the global certainties of the Cold War period and began to doubt the possibility of a better future. In 1977, at a time of social and economic crisis in Britain, the Sex Pistols had pithily summarized the mood as 'No Future'. These changes were accelerated by the beginnings of the era of personal computing that transformed the mysterious world of cybernetics, as computer operations had been known, into a space for individual exploration, named in 1984 by science fiction writer William Gibson as 'cyberspace'. Visual culture burst onto the academic scene at that time, mixing feminist and political criticism of high art with the study of popular culture and the new digital image.

Today there is a new world-view being produced by people making, watching and circulating images in quantities and ways that could never have been anticipated in 1990. Visual culture is now the study of how to understand change in a world too enormous to see but vital to imagine. A vast new range of books, courses, degrees, exhibitions and even museums all propose to examine this emerging transformation. The difference between the concept of visual culture in 1990 and the one we have today is the difference between seeing something in a specific viewing space, such as a museum or a cinema, and in the image-dominated network society. In 1990, you had to go to a cinema to see films (except reruns on TV), to an art gallery to see art, or visit someone's house to see their photographs. Now of course we do all that online and moreover, whenever we happen to choose to do so. Networks have redistributed and expanded the viewing space, while often contracting the size of the screen on which images are viewed, and deteriorating their quality. Visual culture today is the key manifestation in everyday life of what sociologist Manuel Castells calls 'the network society', a way of social life that takes its shape from electronic information networks (1996). It is not just that networks give us access to images – the image relates to networked life on- and offline and the ways we think about and experience those relations.

Simply put, the question at stake for visual culture is, then, how to see the world? More precisely, it involves how to see the world in a time of dynamic change and vastly expanded quantities of imagery, implying many different points of view. The world we live in now is not the same as it was just five years ago. Of course, this has always been true to some extent. But more has changed and changed more quickly than ever and, because of the global network society, change in one location now matters everywhere.

Rather than try to summarize the immense quantity of visual information available, this book offers a toolkit for thinking about visual culture. Its way of seeing the world centres on the following ideas:

- All media are social media. We use them to depict ourselves to others.
- Seeing is actually a system of sensory feedback from the whole body, not just the eyes.
- Visualizing, by contrast, uses airborne technology to depict the world as a space for war.
- Our bodies are now extensions of data networks, clicking, linking and taking selfies.
- We render what we see and understand on screens that go everywhere with us.
- This understanding is the result of a mixture of seeing and learning not to see.
- Visual culture is something we engage in as an active way to create change, not just a way to see what is happening.

While the present day is the focus, much of this book is none-theless historical, as it traces the roots of visual culture today, both as a field of study and a fact of everyday life. The emphasis is no longer on the medium or the message, with apologies to Marshall McLuhan (1964). Instead, the emphasis is on creating and exploring new archives of visual materials, mapping them to discover connections between what is visual and the culture as a whole, and realizing that what we are learning to see above all is change on the global scale.

The book begins by looking at the evolution of the self-portrait into the omnipresent selfie. The selfie is the first visual product of the new networked, urban global youth culture. Because the selfie draws on the history of the self-portrait, it will also allow us to explore the creation of the academic discipline of visual culture that emerged around 1990. How we see ourselves leads to the question of how we see, and the remarkable insights of neuroscience (Chapter 2). Human vision now seems like the multi-faceted feedback loop that visual artists and visual culture scholars have long assumed it to be. Seeing is not believing. It is something we do, a kind of performance. What this performance is to everyday life, 'visualizing' is to war (Chapter 3). Battlefields were visualized first in the mind's eye of the general and then from the air by balloons, aircraft, satellites and now drones. These views of the world are not experienced directly but on screens. So Chapter 4 looks at two examples of the creation of networked worlds: the view seen from a train and the creation of motion pictures; and today's ubiquitous networked digital screens. Those screens appear to offer unlimited freedom but are carefully controlled and filtered views of the world.

The key places in these networks are the global cities, where most of us now live (Chapter 5). In these immense, dense spaces, we learn how to see – and also not to see potentially disturbing sights – as a condition for daily survival. Global cities have grown up around the remains of the imperial and divided Cold War cities that preceded them. They are spaces of erasure, ghosts and fakes. The creation of the global city world has come at tremendous cost. Now we have to learn how to see the changing natural world (Chapter 6). Or more exactly, we have to become aware of how humans have turned the planet into one enormous human artefact, the largest work of art ever made or ever possible.

At the same time, the global city has also become rebellious, the site of permanent unrest (Chapter 7). Here the youthful majority in cities use their connections to claim new ways to represent themselves on social media that are transforming what politics means, from the city revolts in the developing world, such as those in Cairo, Kiev and Hong Kong, to the separatist movements in the developed world, from Scotland to Catalonia. Do we live in cities? Or regions? Or nations? Or power blocks like the European Union? How do we see the place where we live in the world?

The time of change

Though the transformations of the present may appear unprecedented, there have been many similar periods of dramatic change in the visible world before. The nineteenth century was famously described by the historian Jean-Louis Comolli as a 'frenzy of the visible' because of the invention of photography, film, X-ray and many other now forgotten visual technologies in the period (Comolli 1980). The development of maps, microscopes, telescopes and other devices made the seventeenth century another era of visual discovery in Europe. And so we could continue back to the first cosmographic representation of the world on a clay tablet from 2500 BCE. But the transformation of the visual image since the rise of personal computing and the Internet is different in terms of sheer quantity, geographic extent and its convergence on the digital.

If we look in a longer historical perspective, we can perceive the extraordinary pace of change. The first moving images were recorded by the Lumière brothers in France in

1895. A little more than a century later, the moving image has become astonishingly widespread and easily available. The first available video cameras for personal use appeared only in 1985. They were heavy, shoulder-borne devices and not well suited for casual use. It was not until the invention of digital videotape in 1995 that home video became a practical possibility. Editing was still an expensive and difficult proposition until the introduction of programs like Apple's iMovie in 2000. And now you can shoot and edit HD video on your phone and post it to the Internet. Above and beyond personal possession, far more people can see and share all this material via the Internet, the first truly global medium. More people still have access to television but hardly anyone has influence over what is shown on television and fewer still can place their own work on TV. By the end of the decade, the Internet will change how we look at everything, including how we see the world.

To understand the difference, we can compare the distribution and circulation of printed matter. In 2011, according to UNESCO, there were over 2.2 million books published. The last European who was thought to have read all available printed books was the sixteenth-century reformer Erasmus (1466–1536). Over the long lifetime of print, many other means of getting published have emerged, from the letter to the editor to self-produced pamphlets and photocopied documents. The book has still remained the format most likely to convince and impress. However, book publishing is open only to authors who can convince editors to produce their work. Now, the Internet allows everyone with a connection to disseminate their writing in ways that are not visibly different from those used by formal book publishers. The global success of E. L. James's self-published novel *Fifty Shades of Grey*, which has sold over 100 million copies in its Random House reprint, would not have been imaginable even a decade ago. The transformation of visual images, especially moving images, has been still faster and more extensive.

The change at hand is not simply one of quantity but of kind. All the 'images', whether moving or still, that appear in the new archives are variants of digital information. Technically, they are not 'images' at all, but rendered results of computation. As digital scholar Wendy Hui Kyong Chun puts it, 'when the computer does let us "see" what we cannot normally see, or even when it acts like a transparent medium through video chat, it does not simply relay what is on the other side: it computes' (Chun 2011). When an ultrasound scanner measures the inside of a person's body using sound waves, the machine computes the result in digital format and renders it as what we take to be an image. But it is only a computation. A modern camera still makes a shutter sound when you press the button, but the mirror that used to move, making that noise, is no longer there. The digital camera references the analogue film camera without being the same. In many cases, what we can 'see' in the image, we could never see with our own eyes. What we see in the photograph is a computation, itself created by 'tiling' different images that were further processed to generate colour and contrast. It is a way to see the world enabled by machines.

Analogue photographs were certainly also manipulated, whether by editing or darkroom derived techniques.

Nonetheless, there was some form of light source, impacting a light-sensitive surface that we can work out from the resulting photograph. A digital image is a computed rendition of digital input, derived from the camera's sensor. So it is much easier and faster to alter the result, especially now that programs such as Instagram will create effects at a single click. Some of these effects imitate specific formats, like black-and-white film or Polaroid. Others mimic skilled techniques that would have been used in the darkroom when developing film.

Early in the digital era, some were concerned that we would not be able to tell whether digital images had been manipulated or not. It turns out that at both amateur and professional level, it is often not that hard to detect. For example, most magazine readers now assume that all photographs of models and celebrities have been adjusted. Readers operate a flexible zone of viewing, in which it is accepted that a photograph can be altered but not changed so much that it's absurd. Some advertising campaigns now even celebrate their use of 'real' models, knowing that we understand ordinary advertising photographs are manipulated. At the technical level, a skilled user can tell not only if an image has been manipulated, but how and with what program. In early 2013, a star college American football player named Manti Te'o was discovered to have created a story regarding the death of a fake girlfriend to gain sympathy and attention. Once web users were alerted to this possibility, it took less than 24 hours for them to reverse research the photograph he had circulated and discover it was not the woman he claimed. There are websites devoted to reverse search now. Previously, it would have required a detective to do in days or weeks what can be done in seconds with a few clicks.

At the time of the *Apollo 17* mission in 1972, the British art historian John Berger made a brilliant television series and an accompanying book for the BBC called *Ways of Seeing*. The immense success of both projects put the concept of the image into popular circulation. Berger defined the image as 'a sight which has been recreated or reproduced' (1973). He flattened the hierarchy of the arts by making a painting or sculpture equivalent in this sense to a photograph or an advertisement. Berger's insight was central to the formation of the concept of visual culture. An influential definition of visual culture in the 1990s was simply 'a history of images' (Bryson, Holly and Moxey 1994). Berger had himself been taking a cue from the German critic Walter Benjamin, whose famous 1936 essay 'The Work of Art in the Age of Mechanical Reproduction' had just been translated into English (1968). Benjamin argued that photography destroyed the idea of the unique image because – at least in theory – infinite and identical copies of any photograph could now be made and distributed. By 1936, this was already old news, because photography was almost a century old. However, new techniques for the mass reproduction of high-quality photographs in magazines and books, as well as the rise of the 'talkies', or films with sound, convinced Benjamin that a new era was at hand.

With the astonishing rise of digital images and imaging, it surely seems that we are experiencing another such moment. The 'image' is now created, or more precisely computed, independently of any sight that might precede it. We continue to call what we see pictures or images, but they are qualitatively different from their predecessors. An analogue photograph is a print created from a negative, every molecule of which has reacted to light. Even the highest-resolution digital photograph is a sampling of what hits the sensor rendered into computer language and computed into something we can see.

Furthermore, what we are experiencing with the Internet is the first truly collective medium, a media commons if you like. It makes no sense to think of the web as a purely individual resource. You might paint and not show anyone the results. If you put something online, you want people to engage with it. Digital commentator Clay Shirky has borrowed a phrase from the novelist James Joyce to capture the result: 'Here comes everybody' (2008). The point here is not simply the scale of the digital commons, impressive though that is. It is certainly not always the quality of the results, which are highly variable. It is the open nature of the experiment.

And that is why, despite the endless junk, the Internet matters. There is a new 'us' on the Internet, and using the Internet, that is different from any 'us' that print culture or media culture has seen before. Anthropologist Benedict Anderson described the 'imagined communities' created by print culture so that the readers of a specific newspaper would come to feel they had something in common (1991). Above all, Anderson stressed how nations came into being as these imagined communities, with powerful and important results. Trying to understand the imaged and imagined communities created by global forms of experience is similarly central to visual culture. The new communities that are emerging on- and offline are not always nations, although they are often nationalist. From the new feminisms to the idea of the 99 percent, people are reimagining how they belong and what that looks like.

What all moments of visual culture have in common is that the 'image' gives a visible form to time and thereby to change. In the eighteenth century, natural historians investigating fossils and sedimentary rocks made the startling discovery that the Earth was far older than the six thousand years of the biblical account (Rudwick 2005). Naturalists began to calculate how many thousands and millions of years were involved. Geologists now refer to this as 'deep time', a time whose scale is vast by comparison with a brief human lifespan but is not infinite. From this perspective, it makes sense that one of the first photographs ever taken by Louis Daguerre in 1839 depicted fossils.

Figure 4 — Daguerre, *Untitled (Shells and Fossils)*

Of course, the fossils remained conveniently still for the camera. More importantly, they were crucial to nineteenth-century debates in natural history, following French scientist Georges Cuvier's insight that fossils revealed past extinctions (1808). Fossils became central to the long drama that culminated with Charles Darwin's *Origin of Species* (1859) concerning the age of the Earth. Was the planet, as certain Christian authorities insisted, only six thousand years old? Or did fossils show that it was many million years old? A photograph is defined by the length of time the light-sensitive medium, whether film or a digital sensor, is exposed to light. As soon as the shutter closes, that instant is past time. The brief exposure of Daguerre's shutter contrasted dramatically with the millennia of geological time and revealed the new human power to save specific instants of time.

Soon, the demands of the new industrial economy forced a second change to time. Time had usually been decided locally in relation to the sun, meaning that cities or towns a few hundred kilometres apart would use different time. The difference did not matter until it became necessary to calculate how trains would cover long distances to a timetable. The 'absolute' time that we still use, designated in highly specific time zones, was created so as to make such calibrations of time and space possible.

In 1840 the Great Western Railway in England was the first to apply this standardized time. A few years later, the painter J. M. W. Turner gave dramatic visual form to the changes in his stunning 1844 canvas *Rain, Steam and Speed: The Great Western Railway*. The train rushes towards

Figure 5 — Turner, *Rain, Steam and Speed*

us, although our viewpoint seems to be suspended in mid-air. The new train, using a modern bridge, has changed time and speed for the first time since the domestication of the horse. It seems to emerge from the swirling rain as if from primeval creation, an earlier subject in Turner's work. A frightened hare running across the tracks (hard to see in reproduction) symbolizes the overtaken forms of natural speed. Overtaken also was painting as the most advanced form of modern visual representation. For all Turner's brilliance, his painting took weeks to make. A photograph can change the world in seconds.

Just a few years later, in 1848, a remarkable daguerreotype of the Chartist meeting on Kennington Common, London, was taken by William Kilburn. The Chartists demanded a new form of political representation, in which every man (not yet woman) over twenty-one could vote and

Figure 6 — Kilburn, *Chartists at Kennington Common*

anyone could be a member of Parliament, regardless of personal wealth. They wanted annual Parliaments to reduce the possibilities of corruption. The rally was called to mark their delivery of a petition to Parliament with what they claimed were five million signatures endorsing these goals. In less than a decade from Daguerre's fossils, the industrial world had transformed the organization and representation of time and space by means of the new time zones and photography. These changes created a desire for a different system of political representation, a subject perfectly well suited to the new visual medium.

We are in another such moment of transformation. Events can be seen as they happen via the Internet, from a dense variety of amateur and professional perspectives, in blogs, magazines, newspapers and social media, using still and moving images of all kinds. The gain in information is offset by the digitally enabled 24/7 work environment for professionals worldwide, while the Chinese workers who produce the digital equipment that makes the new work regime possible are themselves expected to work 11-hour days, plus overtime if required, with an average of one day off a month. The long struggle to limit the working day has been soundly defeated. Time-based media are newly ascendant, creating millions upon millions of slices of time, which we call photographs or videos, in what seem to be ever-shrinking formats like the six-second-long Vine. The obsession with time-based media from photography in the nineteenth century to today's ubiquitous still-and-moving image cameras is the attempt to try and capture change itself.

In 2010 the artist Christian Marclay made an extraordinary installation called *The Clock*. It was a 24-hour montage of clips from films all telling or showing the time so that *The Clock* was itself a chronometer. The very fact that it was possible to make such an immense montage of clips about time indicated that modern visual media are time-based. We date a painting to the specific year it was finished but it is impossible to tell how long it took to paint. A photograph was always of one instant that may or not be known precisely. Today, digital media are always 'time-stamped' as part of their metadata, even if that time is not visibly recorded in the image. At least for now, in the ever-changing present that is the hallmark of urban global spaces, it seems that we use time-based media as a way of both recording and relieving our anxiety over time itself.

In all this speeding up – from the introduction of the railways to the Internet – we have burned in a matter of two centuries, and especially the past thirty-five years, the remnants of millions of years of organic matter that had become fossil fuel. This vaporizing of millennia has now caused the undoing of the hitherto infinitely slow rhythm of deep time itself. What once took centuries, even millennia, happens in a single human lifetime. As the ice-caps melt, gases that were frozen hundreds of thousands of years ago are released into the atmosphere. You can say that time travel is as simple as breathing these days, at least at the molecular level. The entire planetary system, from the rocks to the highest atmosphere, is out of joint and will remain so for longer than hitherto existing human history, even if we stop all emissions tomorrow.

Where does all this lead? It is too early to tell. When the printing press was invented, it was not possible to imagine from the first publications how mass literacy would change the world. In the past two centuries, the elite military skill of visualizing, which imagined how battlefields that were too large to see with the naked eye 'looked', has been transformed into the visual culture of hundreds of millions. It is confusing, anarchic, liberating and worrying all at once. In the chapters that follow, *How to See the World* will suggest how we can organize and make sense of these changes to our visual world. We will see what is on the rise, what is falling back and what is being strongly contested. Unlike the Apollo astronauts, we will have our feet firmly on the ground. But there is more to see than they could have imagined.

How to See Yourself

In 2013, the *Oxford English Dictionary* announced that its word of the year was 'selfie', which it defined as 'a photograph that one has taken of oneself, typically one taken with a smartphone or webcam and uploaded to a social media website'. Apparently, the word was used 17,000 percent more often between October 2012 and October 2013 than the previous year, due in part to the popularity of the mobile photo-sharing site Instagram. In 2013, 184 million pictures were tagged as selfies on Instagram alone. The selfie is a striking example of how once elite pursuits have become a global visual culture. At one time, self-portraits were the preserve of a highly skilled few. Now anyone with a camera phone can make one.

The selfie resonates not because it is new, but because it expresses, develops, expands and intensifies the long history of the self-portrait. The self-portrait showed to others the status of the person depicted. In this sense, what we have come to call our own 'image' – the interface of the way we think we look and the way others see us – is the first and fundamental object of global visual culture. The selfie depicts the drama of our own daily performance of ourselves in tension with our inner emotions that may or may not be expressed as we wish. At each stage of the self-portrait's expansion, more and more people have been able to depict themselves. Today's young, urban, networked majority has reworked the history of the self-portrait to make the selfie into the first visual signature of the new era.

For most of the modern era, the possibility of seeing an image of oneself was limited to the wealthy and the powerful. The invention of photography in 1839 soon led to the development of cheap photographic formats that placed the portrait and the self-portrait in the reach of most working people in industrialized nations. In 2013, these two histories converged. At the funeral of Nelson Mandela on December 10 that year, Danish prime minister Helle Thorning-Schmidt took a selfie that included President Barack Obama and Prime Minister David Cameron.

While some commentators questioned the propriety of the moment, it marked a departure from the lifeless posed

Figure 7 — Thorning-Schmidt, Obama and Cameron taking a selfie

official photograph and a new investment in a popular format. The photograph of the selfie being taken was reprinted worldwide, although the selfie itself was not released to the media. Only a few weeks later, the world's best-known actors converged around Ellen Degeneres at the 2014 Academy Awards to be in a selfie taken by Bradley Cooper that became the most popular tweet to date (also cited as the most popular 'of all time'). The selfie is a fusion of the self-image, the self-portrait of the artist as a hero and the machine image of modern art that works as a digital performance. It has created a new way to think of the history of visual culture as that of the self-portrait.

The imperial self

These intersections – of the self-portrait, the machine image and the digital – have their sources in the history of art, which we can follow. The Spanish painter Diego Velázquez's masterpiece *Las Meninas* (1656) linked the aura of majesty to that of the self-portrait. The painting is a set of visual puns, plays and performances that revolve around the self-portrait of the artist.

As we look at the painting, Velázquez stands to our left-hand side, holding his brushes. The canvas he is working on blocks our view. In the foreground we see the Maids of the title, the curtseying women, who are the attendants of the little girl in white. She is a princess, known as the Infanta, daughter of Philip IV of Spain. At once we notice that almost everyone in the painting is looking at someone or something, which appears to be located at the viewer's vantage

Figure 8 — Velázquez, *Las Meninas*

point. As we look back into the painting, we see two figures in a frame on the wall behind the main group. The frame is much brighter than the other gloomy paintings hanging on the wall and we conclude that it must be a mirror. In fact, is it not reflecting the people everyone is looking at? And these are no ordinary people. They are the King and Queen, which is why everyone seems frozen to the spot.

In a famous analysis of the painting in his book *The Order of Things*, the French philosopher Michel Foucault described it as depicting not just what could be seen within it but the very means of ordering and representing a society ([1966] 1970). The subject of the portrait is the ways in which it is possible to depict living things in a hierarchy depending on the presence of the King, ranging from the dog at the front, to the 'dwarf' who was a court jester, the ladies-in-waiting and other nobility, the painter and the royal presence. Foucault's approach in turn helped inspire what was called the 'new art history', and later, the concept of visual culture. Foucault showed that the place that everyone is looking at is the centre because the King is there, noting:

> the triple function it fulfils in relation to the picture. For in it there occurs an exact superimposition of the model's gaze as it is being painted, of the spectator's as he contemplates the painting, and of the painter's as he is composing his picture.[1]

The mirror reflects the models that the painted painter is working on. It also makes visible by implication the place from which the real Velázquez worked. And it is the same place where we now stand to look at the finished painting. Foucault observes:

> That space where the king and his wife hold sway belongs equally well to the artist and to the spectator: in the depths of the mirror there could also appear – there ought to appear – the anonymous face of the passer-by and that of Velázquez.[2]

So the 'mirror' does not obey the laws of optics so much as it does the laws of Majesty, like the painting itself. The seventeenth century was a period in which monarchs around Europe claimed the power of Absolutism. That is to say, they were more than just people. Kings were God's representatives on earth, symbolized by their being anointed like a priest during the coronation ceremony. Combining secular and spiritual power, the Absolutist monarchs claimed overwhelming power that was centred in their very person.

How, then, should the king be shown to convey a sense of this power? Not every individual person that happened to become a king or queen was impressive. Even the most powerful have their moments of weakness, illness and decline. Against the fallible individual person of the king, European royalty devised the concept known as 'the body of the king', which we can call Majesty. Majesty does not sleep, get ill or become old. It is visualized, not seen. Any action that diminished Majesty was a crime called *lèse-majesté*, violating majesty, which could be severely punished. It even became a criminal offence to take a piece of paper with the king's name on it and crumple it up. Physical attacks on the monarch were met with truly spectacular punishments because it was a double attack on the person of the king or queen and the institution of Majesty.

Las Meninas is invested throughout with this power, making the image of the king at least the equal, and in some ways the superior, of the king himself. It also makes a set of claims for the power of the artist by association. As we have seen, the 'mirror' is not optically accurate. Art historians like Joel Snyder have shown that the arrangement of perspective in the painting does not in fact converge on the mirror but on the arm of the man standing in the open doorway to its right as we look at it (1985). Although the scene appears to show a mirror reflecting the king, it actually shows the mirror reflecting Velázquez's painting of King Philip. It is possible that Velázquez's perspective was not so precise or that he wanted to create a visual trap for his audience. Whatever you believe, the 'mirror' shows something that the spectator would not usually be able to see – either the painting that the artist is working on, as Snyder has it, or the King and Queen standing in front of it, as Foucault had it.

So the mirror misrepresents, but it also shows a world of possibility. *Las Meninas* makes a tremendous claim for the power of the artist, both literally and metaphorically. The remarkable skill of the piece makes it clear that the painter is capable of accomplishments others are not. Only twenty years earlier, Velázquez had to pay the same kind of tax on his art that a shoemaker did on their shoes. Here Velázquez claims the power of Majesty for art by association and by depiction. He also put a red cross on his costume, indicating his claim to the status of nobility, before he could actually claim to be a noble in real life. Today, when it is common to see paintings sell for millions, even hundreds of millions, the elite status of the artist is taken for granted. It is in fact a relatively new and unusual idea that arose first in the imperial nations of the modern world.

Las Meninas plays with what we can see and what we cannot. It keeps out of sight the source of the Spanish monarchy's power and authority, namely its empire in the Americas. Louis XIV (1638–1715), the Absolutist king of France, who married the older half-sister of the Infanta seen in *Las Meninas*, had an obsidian mirror in his Cabinet of Wonders, said to have been plundered from Moctezuma II himself, the last Aztec emperor (ruled 1502–20). Obsidian is a material formed by cooled lava that is both black and reflective. Mexican artist Pedro Lasch, who has worked with the black mirror, emphasizes that 'In pre-Columbian America, as in many other cultures, black mirrors were commonly used for divination . . . The Aztecs directly associated obsidian with Tezcatlipoca, the deadly god of war, sorcery, and sexual transgression.'[3] If the European mirror image was a place of power, its American equivalent added violence, sexual ambivalence and storytelling to the imperial mix.

In both the pre-encounter Americas and in medieval Europe, the mirror was a place of divination, where fortunes were told and where contact could be made with the dead and other spirits. In short, the mirror is a visual bridge between past, present and future.

Figure 9 — Lasch, *Liquid Abstraction*

The imperial portrait in the Absolutist era (1600–1800) was, then, never just one image. The portrait of the individual who happened to be king also depicted the Majesty of the King, or the power of representation itself. The self-portrait of the artist claimed that art was the work of nobility not artisans. The mirror reflects either the real king and queen or the painted portrait of the King. Or, in some not quite mathematical but nonetheless perfectly intelligible sense, both. The black mirror and the optically incorrect painted mirror show us how things are now, but are also a place to access the past and the future. These reflections and images were a combination of theatre, magic, self-fashioning and propaganda that were key to sustaining royal power.

The portrait and the hero

When the old monarchies collapsed during what can be seen as the long age of Revolution (1776–1917), a new 'frenzy of the visible' accompanied and was part of the social transformation (Comolli 1980). Across this era, dramatic inventions of new media like lithography, and especially the various processes we call photography, portraits and self-portraits, seemed to revolutionize the visible. Visual media were democratized. Until this time, the ordinary person might have seen visual images in church, on coins, at parades or in carnivals. By the mid-nineteenth century, there were new museums of art; illustrated newspapers and magazines were being published; and visiting-card photographs could be bought cheaply. New ways of being came to be imagined and visually represented, including the modern artistic 'genius',

Figure 10 — Vigée-Lebrun, *Marie Antoinette*

nearly always male, but also the woman artist. The heroic artist took some of the aura of the King (or Queen) and transferred it to him- or herself. Brought down to earth, the self-portrait became the picture of a hero.

In the last years of Absolutism, the new order was already emerging. The royal artist Élisabeth Vigée-Lebrun painted portraits of the French Queen, Marie Antoinette. She also

Figure 11 — Vigée-Lebrun, *Madame Vigée-Lebrun and Her Daughter Julie*

painted a number of self-portraits. To borrow a cue from John Berger, can you see which is which?

Both women look out at the viewer directly from the painting, against a scumbled background of loosely handled non-representational paint. Both are dressed as fashionable, modern women in the loose style of the period, with their finely handled sashes showing the skill of the artist. Perhaps the informality of the pose with her child allows us to see Vigée-Lebrun in her *Self-Portrait with Her Daughter* (1789). The portrait of Marie Antoinette (1783) became the subject of a scandal precisely because of its informality. At the same time, by so blurring the difference between the Queen and the artist, Vigée-Lebrun claimed a new level of equivalence between the two.

In their classic study *Old Mistresses* – the title is a pun on the phrase 'Old Masters', used to mean distinguished artists of the past with the implication that such artists would be men – Rozsika Parker and Griselda Pollock studied the history of women artists (1981). The self-portrait with her daughter raised particular issues because women were not even supposed to be artists, according to the received prejudice, so a painting by a woman showing a woman artist was doubly defiant. Parker and Pollock described how in Vigée-Lebrun's *Self-Portrait*:

> The novelty [of the painting] lies in the secular and familial emphasis, the Madonna and Child of traditional iconography replaced by mother and *female* child locked in an affectionate embrace. This portrait of the artist and her daughter elaborates that notion of woman, emphasizing that she is a mother.[4]

Vigée-Lebrun had taken the Christian image of the Virgin Mary and Infant Jesus and given it a secular and contemporary spin. Notably, both the artist and her daughter look out at us confidently, unlike the traditional downcast glance of the Madonna in paintings by artists like Raphael. Still, as Parker and Pollock pointed out, there was a Catch-22 here.

In celebrating her role as a mother, unusual in the period when women would often leave their children with wet-nurses, Vigée-Lebrun's picture seems from our perspective like a cliché. The restrictive doctrine of the woman as the domestic angel by the hearth, caring for children but not active professionally, was actually a creation of the nineteenth century. For modern feminists, trying to escape what Betty Friedan famously called 'the feminine mystique' (1963), Lebrun at first looked like more of the same. It took Parker and Pollock's close attention to context and detail to see her work differently.

If the nineteenth century visualized women as domestic help-meets, their counterpart was the idealized Great Man, or hero, as imagined by the historian Thomas Carlyle. For Carlyle, writing in 1840, 'great men make history' (Carlyle 1840). Artists also conceived of themselves as heroes in different ways. What did the modern artist hero look like? In 1839, Louis Daguerre in France and William Henry Fox Talbot in Britain finally produced photographs that 'fixed', meaning that the light-sensitive surface stayed as a visible image, rather than blacking out. Another French practitioner, Hippolyte Bayard, also invented a photographic process at this time. Doomed to the margins of photographic history because his colleague Daguerre was credited with the invention, Bayard nonetheless might be credited with inventing the selfie in his *Self-Portrait as a Drowned Man* (1839–40). He also invented the photographic fake because he was not, of course, actually dead.

Like many a Romantic hero before him, following the example of the poet Goethe's hero Werther who committed

Figure 12 — Bayard, *Self-portrait as a Drowned Man*

suicide in Goethe's enormously successful 1774 novel *The Sorrows of Young Werther*, Bayard pretended to prefer death over dishonour. His photograph is what the writer Ariella Azoulay has called an 'event' (2008). It presupposes that the community watching it can imagine the heroic narrative of the author's suicide and understand his disappointment. Some people even thought that Bayard really was dead and discussed how the dark skin on his hands and face was the consequence of drowning, rather than of exposure to the sun.

The painter Gustave Courbet appropriated the idea of the artist's suicide for his own self-portrait *The Wounded Man* (1845–54). As he too was living in Paris at the time, it is quite possible that he saw or heard about Bayard's photograph.

Figure 13 — Courbet, *The Wounded Man*

Here the artist has apparently stabbed himself but found time to put the sword back up against the tree behind him. Of course, we are no more supposed to think of this painting in this realistic way than we are the photograph. Marshall McLuhan later suggested that new media take the content of old media, such as television adapting theatrical plays to create TV drama (1964). Here, though, the new medium seems to have influenced the old. Courbet had moved from rural France to Paris just in time for the multiple revolutions of 1848, which he supported. By 1855, the revolution had failed and suicide was perhaps the only option left to the true revolutionary. Courbet issued a manifesto at his one-man exhibition that year, declaring: 'To know in order to be able to do, that was my idea'. In this view, painting, like photography, depicts knowledge and leads to action, or the event. For Bayard and Courbet alike, the artist was the hero, the person capable of creating an event, even at the (fictional) cost of their own life.

It's a seductive idea. Writing in the aftermath of the revolutions of 1968, the art historian T. J. Clark used Courbet as his example of 'a time when political art and popular art seemed feasible'. He stressed both Courbet's involvement with politics and the influence of popular media on his painting, arguing that popular art shows 'the essentials of a social situation'. As with Parker and Pollock's work, his ideas are so accepted now that it is hard to appreciate how innovative the approach was in 1973 when his book *The Image of the People* was first published. Art historians began to look at popular prints, photographs and other mass-produced visual material alongside painting and sculpture, a means of research known as social art history. For two decades afterwards, social art history and visual culture studies worked closely together before visual culture became a separate area of study around 1990, largely due to the rise of digital media.

Another reason that division occurred was the increasing difficulty of deciding what was essential, to use Clark's term, in a given moment. The transformation of the arts and humanities since 1968 has been the result of a succession of groups pointing out that they have been overlooked and that their interests need to be taken into account. And then people look back into the historical record and discover that this group was there all along. One example is a self-portrait by the French Impressionist Henri de Toulouse-Lautrec,

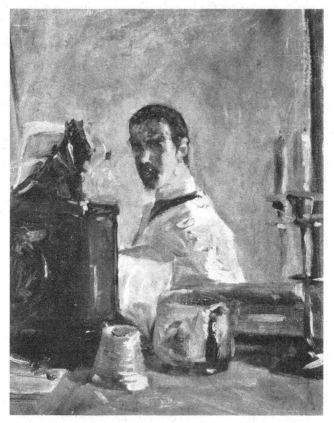

Figure 14 — Toulouse-Lautrec, *Self-Portrait Before a Mirror*

usually remembered for his depictions of Paris night life. Another way to understand his work would be as an artist with disabilities. His *Self-Portrait Before a Mirror* (1882) challenges the conventions of the genre and its interpretation.

Toulouse-Lautrec deliberately painted his reflection in a mirror, rather than just using a mirror to make a self-portrait as was traditional. The reflected candlestick removes any doubt as to whether the frame indicates a mirror or a window, as in the Velázquez, for Toulouse-Lautrec clearly wanted us to recognize it as such.

At the same time, the painting both conceals and reveals the artist. By using a mirror standing on a mantelpiece, he shows us only his head and shoulders. He might have used this device to conceal his disability. For, either as a result of childhood accidents or a congenital condition, Toulouse-Lautrec had an adult upper body but the legs of a child. He depicted himself in the *Self-Portrait* just protruding into the mirror, leaving the top half of the mirror empty, indicating to the observant viewer that he was very short. He might have chosen to adjust what he saw, so as to fill the 'screen', like a present-day actor or politician standing on a riser to seem taller. Whereas for dominant groups the mirror is often a site (and sight) of affirmation, for people who look or feel different, the mirror can be a site of trauma. Toulouse-Lautrec's self-portrait confronts that sight without making himself the object of a freak show. I use the term deliberately because in the period people with disabilities were literally exhibited as 'freaks' to paying audiences (Adams 2001). Toulouse-Lautrec refuses to cater to this voyeuristic desire to see, but does not distort the reality of his difference. It's a different kind of heroism and one that is not immediately recognizable as such to others.

The many selves of postmodernism

In the late 1970s, a new idea began to circulate in European and North American intellectual and artistic circles. The

modern period, defined by its heroic artists, radical political divides and the dramatic expansion of the industrial
economy, seemed to be over. Beginning with thinkers like
the French philosopher Jean-François Lyotard, artists and
writers started to think about a 'postmodern condition'
(1979). At the time, there were two ways of understanding
postmodernism. One view saw it as a break with the modern that could be given a specific date. Another more widespread view held that there had always been a 'postmodern'
side to modernism, which questioned its certainties. The
prime example of a 'postmodern' modern artist was Marcel
Duchamp. He took manufactured objects like a bicycle wheel
or a urinal, installed them in an art gallery or exhibition and
declared the result to be art. In other words, art was whatever someone who wanted to be an artist called art. Whether
it was the product of individual skill or talent was beside the
point. Duchamp called the results 'readymades', perhaps the
best known of which is *Fountain* (1917). It was made from a
urinal, stood on its end and signed 'R. Mutt'. The artist was
no longer a hero.

When Duchamp made *Fountain*, the First World War
had devastated Europe, with millions dead. The Russian
Empire had collapsed into the Revolution that would create the Soviet Union. No wonder Duchamp and other artists thought things had changed. The war had even created a
new set of mental illnesses, for which the term 'shell shock'
was coined, in which sufferers seemed to experience a trau-
matic moment over and over again, or became blind despite
there being no injury to their eyes, and so on. The 'self' no
longer seemed so secure. Perhaps there was more than one

Figure 15 — Duchamp, *Self-Portrait in a Five-Way Mirror*

self in each person. In 1917 Duchamp developed this idea to create a new readymade self-portrait at a store on Broadway in New York City. Using a hinged mirror, the photo booth created a five-way portrait in three copies.

It was a perfect outlet for him. It was visually amusing but had a serious point – Duchamp did not see himself as one but as many selves. Whereas the heroic modern artist simply depicted her or his own image, the postmodern artist makes him- or herself as their primary project. Nor is this a once-and-for-all remake but it can be done over and over. It is not an event but a performance.

Duchamp continued to experiment with his self-image. He collaborated with his friend Man Ray to create a self-portrait as his alter ego Rrose Sélavy. In order to get the pun, you need to read this name with a French accent when it will become 'Eros, c'est la vie', meaning 'love is life'.

As if intent on making the point that Rrose Sélavy was not his 'real' identity, Duchamp made several notably

Figure 16 — Man Ray, *Marcel Duchamp as Rrose Sélavy*

different versions of this self-portrait. The one shown here is perhaps the most feminized, in the style of a society portrait or fashion illustration. The implication of the portrait, like all drag, is that gender is a performance. Like seeing, it is something we do, rather than something that is inherent and unchangeable. Rrose Sélavy seems feminine because of her clothes, make-up and jewellery, as well as the way she holds her body. In her classic study *The Second Sex* (1947), French writer Simone de Beauvoir put it pithily: 'One is not born, but rather becomes, a woman.'[5]

Such public claims to feminist and queer identities as something we do, and can therefore change, combined with the rise of mass personal image-making technologies and personal computing, were central to the creation of the field of visual culture. It is important to recognize how transformative these interfaces were in what has become known as the postmodern period (1977–2001). Far from being simply negative critiques, they inspired some remarkable creative accomplishments, such as the work of the New York artist Cindy Sherman, whose awareness of feminism, combined with her DIY photographic aesthetics, influenced a new generation of artists, writers and academics. In the art world, she was known as part of the Picture Generation. Her work was also key for the study of visual culture.

From her time as a graduate student in Buffalo, NY, Sherman has repeatedly photographed herself in an ever-changing variety of poses and attitudes to explore how we make ourselves and make our gender. In her classic early series called *Untitled Film Stills* (1977–80), now owned by the Museum of Modern Art, New York, Sherman set out to counter the construction of women as passive objects of male desire. In the heyday of Hollywood cinema, film stills were used as a form of publicity for new movies. They would be displayed outside cinemas or used in print, whether as advertising or to illustrate a review. Cinema fans used to collect these stills, in the way that baseball fans collected baseball cards. In 1977, the classic Hollywood studio film already felt dated, so Sherman's work was really about how the then-present wanted to distinguish itself from the time in which women were only to be 'looked at' (Berger 1973). She created a long series of black-and-white photographs of herself in different costumes, make-up and situations, exploring the ways in which cinema looks at women.

Sherman began her project just two years after film critic Laura Mulvey had coined the expression 'the male gaze' in a study of classic Hollywood cinema (1975). Mulvey saw that a gaze (that is, a dominant way of seeing) is built into cinema, which can be that of the actors but is also part of the medium itself. The man's role in the film, Mulvey says, is 'the active one of forwarding the story, making things happen'. She adds that the man in the story 'controls the film fantasy and also emerges as the representative of power in a further sense: as the bearer of the look of the spectator'.[6] Men look at the action through the eyes of the male hero and women are obliged to do the same, a form of compulsory gender manipulation. The cinematic gaze also performs the action in which 'I see myself seeing myself', that sense that we sometimes have of being looked at, even if we can't actually see the person doing the looking. For Mulvey and other feminists, women experience this condition all the time in relation to how they look and act. By freezing the film and making us think precisely about how and why we are looking or being looked at, Sherman made this performance visible.

If we look at this example from 1978, it is clear how skilfully Sherman has put the image together. The low camera angle makes it seem that we see the woman in the picture (who is always Cindy Sherman herself) but that she does

Figure 17 — Sherman, *Untitled Film Still*

not see us. She appears isolated and trapped in the cityscape that hems her in. Using sharp side-lighting and close focus, Sherman makes her body stand out from its surroundings. While this might have been a confident pose if she was looking directly at the camera, she is instead looking away to the side at something we cannot see, her lips slightly parted, creating a sense of threat and anxiety. In the classic Hollywood *mise-en-scène* (the creation of both the individual shot and the overall feel of the film), the victim is always isolated like this before being subjected to violence. At first we can't help but feel anxious on the woman's behalf. Then we realize that it is Sherman herself who has created the scene and that she is using it not to present herself as a victim but to make us aware of the ways in which cinema depicts women as objects to be played with. By manipulating back, Sherman and many other artists of her generation, such as Barbara Kruger and Sherrie Levine, claimed the right to be the selves they wanted to be. Her photographs re-perform the way women are represented to say something important about the actual experience of women in daily life.

Photographic self-portraits can also be a diary and a record of what has happened. In a counter-example to Sherman's role-playing, the New York photographer Nan Goldin kept such a diary over many years. It recorded her radical and alternative counter-culture circle in 1980s New York. Goldin would show her photographs in a darkened room as a slide show, using a carousel slide-projector in those pre-PowerPoint days. Accompanied by a soundtrack of music from the Velvet Underground and other downtown classics, the performances would last for about an hour, immersing the audience in a visual narrative of Goldin's life. Viewers would come to recognize her friends and her boyfriend. So it was a visual shock in 1984 when she created a photograph of herself with visible bruising on her face. A second shock was found in its title: *Nan One Month After Being Battered*.

As damaged as her face appears to be in the photograph, we then realize that she has had a month to recover, so the initial violence must have been dreadful. Goldin warns that we can depict ourselves but it does not always mean we can protect ourselves.

While one strand of postmodern art and thought highlighted the illusions of modern consumer society, work like Goldin's inspired a new generation of artists and writers to concentrate on how gender, race and sexuality were

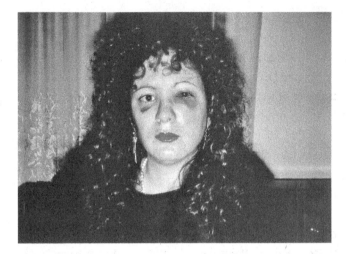

Figure 18 — Goldin, *Nan One Month After Being Battered*

experienced in everyday life. In a word: performance. Performance, in the classic definition of scholar Richard Schechner, is 'twice-performed behavior'.[7] Schechner claimed that all forms of human activity are a performance, assembled from other actions we have taken in the past to create a new whole. A performance might be an artwork, it might be a chef cooking a dish or a barber cutting hair. Or then again it might be anyone whatever giving a performance on their gender, race and sexuality in everyday life.

Putting the other in shade

It was in 1990 that this visual culture of performance became visible in the United States, extending from the avant-garde to academia and the mainstream. First, Jennie Livingston's remarkable documentary *Paris Is Burning* (1990) made the subculture of the queer voguing trend in Harlem available to art-cinema audiences. When the popstar Madonna adopted the style for her hit *Vogue* that same year, especially in its compelling video, the global media audience saw what it meant to 'strike a pose'. In a related vein, the philosopher Judith Butler published her classic book *Gender Trouble*, which showed how drag reveals gender itself to be a performance (1990). And in both the United States and United Kingdom, degrees in visual culture were offered at the University of Rochester and Middlesex University for the first time.

Voguing was a dance form created at the balls organized by gay African-American and Latino men in Harlem. These balls are a mixture of dance and performance in which participants 'walk' in competition for prizes. According to the narrative offered in *Paris Is Burning* by the performer Dorian Corey, balls initially featured only drag queens and later expanded to categories like film and television stars and all manner of categories taken from 'real life'. These last included military personnel, executives and students. All these categories were felt to be ways of being or careers that were desirable but not open to gay men of colour. (Queer had yet to become an affirmative term but it was also in 1990 that the Queer Nation activist group was formed.) The goal was to present 'realness', meaning that if you were out in public away from the ball, you could pass as really being whatever category you were representing. The ball was a mirror to the real world in the sense that it was reversed: here gay African-American and Latino men were in the ascendant, becoming legendary in their houses.

The 1980s saw the rise of voguing at the balls. Voguing is a competitive form of dance, using frozen and exaggerated positions in time to the house music of the period. In the earlier styles, walkers looked to find a 'read' on their rivals, meaning a flaw in their costume or appearance. One read in the film involved a debate as to whether a walker in an upscale male category was using a woman's coat and was therefore disqualified. In the ballroom, to be read was to fail. It is to be seen by the other as they wish to see you, rather than as you see yourself. You wanted to simply appear to be what you appeared to be. In short, for your performance to succeed so well that it becomes invisible as a performance. The read reveals otherwise.

By contrast, a vogue makes you see yourself differently. As voguer Willie Ninja demonstrates in the film, it might involve a mime in which the dancer checks their appearance in a 'mirror' and then holds that (non-existent) mirror up to the opponent to show them how deficient they are by comparison. Ninja's vogue created what was called a 'shade', making you see yourself seeing yourself in the pretend mirror. It is therefore a more devastating move because you are convinced of its truth, whereas you could – and people can be seen in the film doing just this – deny the 'read'. The read and the shade are operations of the gaze but with a difference – or queered, as we might say now. Unlike the male gaze, where maleness is assumed because of genital difference, the assumption of gender here is taken on voluntarily and then performed. Some of the men assume female roles, others male, some vogue. In *Paris Is Burning*, we see what happens when the gaze is both queered and used by people of colour. Like her or not, Madonna's hit song and video *Vogue* brought the ballroom subculture and the possibilities of performing yourself to global public attention.

In *Gender Trouble*, Butler held such drag performances to demonstrate that 'gender is the cultural meaning that the sexed body assumes'. By this she meant that we cannot draw a direct equivalency between a person's gender and their body's sexual organs. Further, she emphasized that bodies do not fall neatly into two 'sexes'. People of intersex constitute some 1.7 percent of live births (in 2013 Germany legislated that people could be designated as 'indeterminate gender' at birth). Butler's point is that even if this is rare, it shows that there is not an absolute equivalency between types of body and gender. Intersex people can make a choice or have one made for them. Drag performers and transgender people can do so in other ways. Judith Halberstam called one such option 'female masculinity', the way that some women deploy masculinity as the cultural meaning of their bodies (1998). If we decide a person's gender by their hair, clothes and style, it is a visual analysis rather than a scientific deduction. As Butler put it, the question becomes: 'What are the categories through which one sees?' Even seeing a naked body might not be enough to decide 'whether the body encountered is that of a man or a woman',[8] and what those decisions mean. Although it is a difficult and serious book, *Gender Trouble* was a crossover hit, as likely to be read in nightclubs as in seminars. It has been part of a transformation of attitudes to gender and sexuality. In so doing, it also helped shape the study of what has come to be called visual culture.

Using the self as a performance that can be photographed has had a variety of dramatic effects. In 1977, in the Central African Republic, a young African photographer named Samuel Fosso was beginning to use left-over film in the photographic studio where he worked to make posed self-portraits. In the same way that Sherman and others have explored how gender is imposed on our bodies from outside, Fosso has visualized how his body is 'Africanized' and 'racialized'.

Figure 19 — Fosso, *The Chief (the one who sold Africa to the colonists)*

This process had been analysed by Frantz Fanon, the Caribbean writer and activist. Sometime in the early 1950s, Fanon was travelling in a French train, headed for his psychiatric training. He described the experience in his book *Black Skin, White Masks* (1952). A child saw him and cried out: 'Look, a Negro! Mama! Look, a Negro! I'm scared'. Fanon recalled how 'the Other fixes you with his gaze, gestures and attitude'. This is a form of photograph, a colonizing power of looking, or, in terms of the ballroom a 'read'. Fanon felt forced to 'cast an objective gaze over myself, [I] discovered my blackness'. He finds himself seeing himself as the white other sees him – a 'shade'. He feels fixed, as if photographed by what he calls 'the white gaze'.[9] Under that gaze, he cannot be seen for himself but only as a set of clichés and stereotypes.

Fosso set out to undo the white gaze by making fun of it in his self-portraits. He has described this particular self-portrait as follows:

> I am an African chief, in a western chair with a leopard-skin cover, and a bouquet of sunflowers. I am all the African chiefs who have sold their continent to the white men. I am saying: we had our own systems, our own rulers, before you came. It's about the history of the white man and the black man in Africa.[10]

Fosso alluded to the fact that the tribal system, dominated by chiefs, was a creation of colonial powers, rather than being 'traditionally' African. As we can still see today, colonial powers have preferred to rule through intermediaries. Often, these people had no legitimacy of their own and so they relied on the colonizer's authority and armies. Mobuto Sese Soko, the former dictator of Zaire from 1965 to 1997, made the leopard skin that we see in Fosso's photograph into a visual cliché of this kind. Zaire was the name Mobuto gave to the former Belgian Congo but his supposedly authentic leopard-skin caps were actually made in France. For all his rhetoric of authenticity, Mobuto's regime was really enabled by the Cold War. The United States tolerated his abuses because he was anti-communist at a time when Africa was far more sympathetic to socialism than capitalism. Fosso wants us to see that even after the end of the formal colonial regimes, Africa is still shaped by them, even as he ridicules such puppet leaders.

Selfies and the planetary majority

In the present moment of transformation, these categories of identity are being remade and reshaped. Today, claims queer theorist Jack Halberstam, 'the building blocks of human identity imagined and cemented in the last century – what we call gender, sex, race and class – have changed so radically that new life can be glimpsed ahead'.[11] One place where we can catch sight of these glimpses is the selfie. When ordinary people pose themselves in the most flattering way they can, they take over the role of artist-as-hero. Each selfie is a performance of a person as they hope to be seen by others. The selfie adopted the machine-made aesthetic of postmodernism and then adapted it for the worldwide Internet audience. It is both online and in our real-world interactions with technology that we experience today's new visual culture. Our bodies are now in the network and in the world at the same time.

Some see the new digital performance culture as self-obsessed and tacky. It is more important to recognize that it is new. The only thing that we know for sure about the young urban global network is that it will change frequently and unpredictably, using formats that may make no sense to older generations. The selfie is a new form of predominantly visual digital conversation at one level. At another level, it is the first format of the new global majority and that is its true importance.

The 'selfie' took off following the placement of a good-quality front camera on the iPhone 4 in 2010, with other phones rapidly following suit. Selfies could now be taken outside or using a flash without the resulting burst of light dominating the picture, as it did in pictures taken in a mirror, which were a staple on the social networking website MySpace in its heyday from 2003 to 2008. A 'selfie' is now understood as a picture of yourself (or including yourself) that you take yourself by holding the camera at arm's length. A set visual vocabulary for the standard selfie has emerged. They look better taken from above with the subject looking up at the camera. The picture usually concentrates on the face, with the risk of making a Duck Face, which involves a prominent pout of the lips. If you overdo it and suck in your cheeks too far, *voilà*, the duck face. These poses are remaking the global self-portrait.

Despite the name, the selfie is really about social groups and communications within those groups. The majority of these pictures are taken by young women, mostly teenagers, and are largely intended to be seen by their friends. In an analysis for the website SelfieCity, media scholar Lev Manovich has shown that – worldwide – women take the majority of selfies, sometimes by overwhelming margins, as in Moscow where women take 82 percent of all selfies (SelfieCity). They are then shared in social circles that are likely to be mostly women, regardless of sexual orientation. As fashion critics have long asserted, (straight) women dress as much for each other as for men and the same can be said of the selfie. Some have suggested that the premium on attractiveness indicates that the selfie is still subject to the male gaze. Sociology professor Ben Agger has claimed in media interviews that the selfie is the male gaze gone viral, part of what he calls 'the dating and mating game'. But trends for #uglyselfies and to show non-conventional selfies are equally apparent. By the nature of the medium, any one person can only see a very limited number of the total selfie production, and even then needs a good deal of extra information to be confident as to what is being seen.

As the format rose to prominence, there was certainly a moral panic in the media about selfies (Agger 2012). A typical comment by CNN commentator Roy Peter Clark declared: 'Maybe the connotation of selfie should be selfish: self-absorbed, narcissistic, the center of our own universe, a hall of mirrors in which each reflection is our own'.[12] In *Esquire*, novelist Stephen Marche went a step further, claiming: 'The selfie is the masturbation of self-image, and I mean that entirely as a compliment. It gives control. It gives release'.[13] These metaphors are slightly convoluted. Narcissus spent his life looking at himself but he did not release a copy of his image for others to look at. Selfies are, like them or not, all about sharing. Many celebrity selfies, like the naked photograph sent out by journalist Geraldo Rivera, have been greeted with scorn. At a private level, a selfie might be liked by some friends but disliked or even satirized by others. This is not masturbation. It's an invitation to others to like or dislike what you have made and to participate in a visual conversation.

Something is happening here, as the numbers suggest. In Britain alone, 35 million selfies were being posted to the Internet each month by 2013. By mid-2014, Google claimed that 93 million selfies were being posted worldwide every day, over 30 billion a year. In her analysis of the photographs on SelfieCity, media scholar Elizabeth Losh found four technical commonalities. First, these pictures are all taken from close distance. You could use a remote but people choose not to do so: the close-up is part of what makes a selfie. The selfie shows that our bodies have become incorporated into the digital network and are interacting with it. To use a remote or timer would be to introduce a distance between the body and the network. As a result, the device that takes the selfie is often visible in the picture. Such mirroring is relatively rare in painting and traditional photography but is not felt to be intrusive in a selfie. By the same token, selfies often use filters like those provided by Instagram that are not designed by the photographer.

Losh sees this 'authoring' by pre-formed tools as taking over from traditional authorship, in which taking decisions as to how an image would be rendered was central. This leads her to conclude that machines are starting to do our seeing for us, using their defaults that we may not understand to shape our perception.[14] It's not altogether new, as we have seen with the Duchamp example. Even in professional contexts, the settings on the Leica camera determined the appearance of classic photojournalism, generating a sharp focus in the foreground and a blurry background. By the same token, the rich colour and depth of field of the current Canon G series has set the visual terms for 'prosumer' photography. The selfie is different by virtue of scale. When Duchamp played with machine vision, it was known to a tiny circle of his associates. The machine vision of the iPhone was used by 500 million people as of March 2014, according to Apple, with a million new phones being sold every three days.

There are really two kinds of 'selfie' in terms of content. One is a performance for your digital circle. A celebrity selfie, like those of Kim Kardashian, is intended to maintain and extend the celebrity of its subject. The celebrity selfie is a continuation of the film still and advertising shot that pretends to be the work of its star. Just as no one who receives a mass email from 'Barack Obama' assumes that the President actually wrote it, the celebrity has not posed at random. Both have undoubtedly some oversight role in the product but it is a controlled form of performance. Far more common, although invisible to those not directly involved, is the selfie as digital conversation, shared via apps like Snapchat.

There are many warnings that the Internet archives material for ever and so a silly, stoned or sexual photograph posted to Facebook could cost you a scholarship or a job.

Figure 20 — Snapchat ad

Although the few documented instances seem to show that people are mostly fired for writing disparaging things about their current job, one poll in 2013 reported that 10 percent of 16–24 year olds claimed to have lost a job because of things they posted online. As a result, many have shifted to using apps like Snapchat for photographs so that Internet users cannot find them once they have been deleted. Once you open a 'snap', you have ten seconds to look at it before it automatically deletes itself. 'Snap' use rose from 200 million a day in June 2013 to 700 million a day by May 2014. That's over 250 billion snaps a year, seen only by the recipients. Users can send snaps to friends of their choice and, unlike email or Facebook, Snapchat also tells you if your friend has looked at the snap or if they have taken a screenshot image. Snapchat's self-image ad (above) reflects its target audience of young women (perhaps unsurprisingly, they are conventionally attractive, white and blonde in this case). The snap they are making is of both of them, presumably intended for their friends.

Snapchats can also convey messages, share information and are designed to sustain conversations. The snap has taken over for many young people from the Facebook status update, just as Facebook pushed out MySpace. The interest for us is not in the specific platform but the development of a new visual conversation medium, usually relayed by phones that are used less and less for verbal exchange. The selfie and the Snap are digital performances of learned visual vocabularies that have built-in possibilities for improvisation and for failure. Networked cultures are intensifying the visual component and moving past speech.

Together with Snapchat has come the Vine, the six-second video message. Vine seems to be the logical outcome of people wanting to get straight to the 'good bit' on YouTube. In six seconds, there's not much time to get bored, you would think. After a while, though, many Vines seem the same – sporting feats, pet and animal tricks, accidents that are supposed to be funny. There are also people using them very creatively as short movies and inevitably corporations have started making ads. Vine has been bought by Twitter, which makes sense. The 140-character message is now supplemented by the six-second video.

Now we see the digital performance of the self becoming a conversation. Visual images are dense with information, allowing successful performances to convey much more than the basic text message, whether in a single image or short video. The selfie and its other forms like Snapchat have given the first visual form to the new global majority's conversation with itself. This conversation is fast, intense and visual. Because it draws on the long history of the self-portrait, it's likely that the selfie in one form or another will continue to play a role in shaping how to see people for a long time to come. The selfie shows how a global visual culture is now a standard part of everyday life for millions that takes the performance of our own 'image' as its point of departure.

How We Think About Seeing

Seeing is something we do, and we continually learn how to do it. It is now clear that modern visual technology is a part of that learning process. Seeing is changing. A widely cited 2006 study from the University of Rochester showed that playing video games improved both peripheral and central visual perception. In other words, playing visual games makes you see better. There are many such reports of improved hand-eye co-ordination. In 2010 another Rochester study showed that gamers make faster and more accurate decisions based on sensory perceptions. Lead author Daphne Bavelier (now at the University of Geneva) describes this as 'probablistic inference', meaning the kinds of decisions we make based on incomplete information, such as choices made while driving (Bavelier Lab). The point here is that we do not actually 'see' with our eyes but with our brains. And we have learned that in turn by becoming able to see how the brain operates. What we see with the eyes, it turns out, is less like a photograph than it is like a rapidly drawn sketch. Seeing the world is not about how we see but about what we make of what we see. We put together an understanding of the world that makes sense from what we already know or think we know.

It has long been realized that we do not see exactly what there is to be seen. The ancient Greek architects of the Parthenon in Athens designed the sides of their columns with a slight outward curve (entasis) as they rose in order to convey the appearance of being perfectly straight. In the seventeenth century, Western science began to distinguish between biological sight, which sees what there is to see, and cultural judgement, which makes sense of it. The philosopher and natural scientist René Descartes pointed out that when we look at a work of art drawn in perspective we perceive what is actually an oval as a circle. He interpreted this as evidence that judgement corrects the perception of sight. This understanding was the basis of modern observational science. Descartes moved the knowledge of the world from being derived from the classical thought of ancient Greece and Rome to what each person observes in his famous aphorism: 'I think, therefore I am' (Descartes 1637). Only the fact that we think indicates that we exist. Everything else must be doubted and tested.

Descartes used vision as his example. The ancient Greeks and Romans had two contradictory theories to explain vision. One said that the eye threw out rays to 'touch' the things we see. The problem with this idea is that we can see very distant objects immediately: so how does vision throw its rays so fast? Another theory said that objects emitted little copies of themselves that got smaller and smaller until they entered the eye. The problem here is that large objects can be seen close up and enormous objects like mountains can also be seen: how did the copies get small enough, quickly enough, to enter the eye? No one could solve these

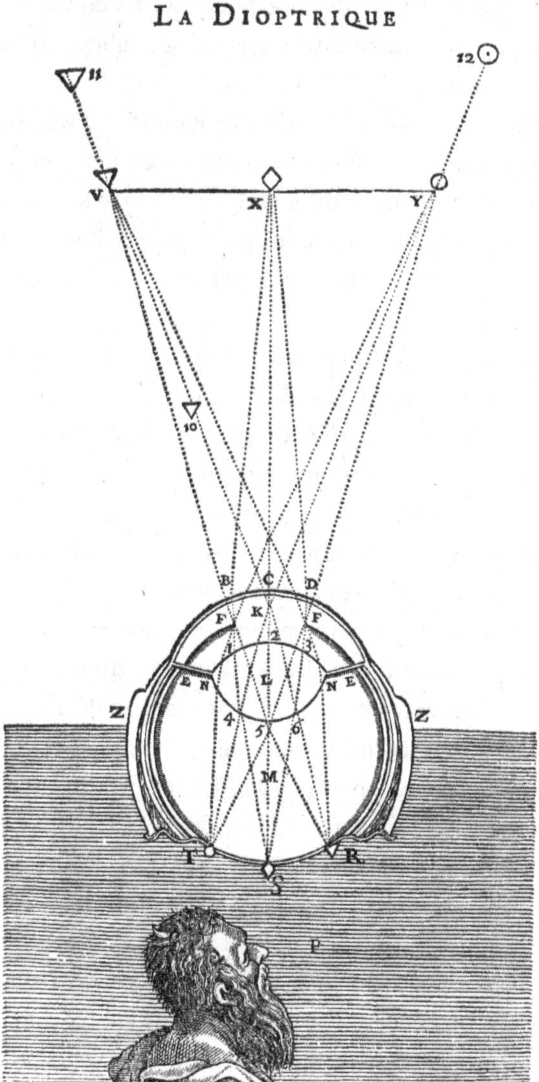

Figure 21 – Descartes, 'Vision', from *La Dioptrique*

problems and they did not really try to do so because light was held to be divine and so not subject to human understanding.

Descartes believed that the existence of God was the only way to guarantee that our observations are not simply delusions or the ravings of the insane. So he tested everything. In 1637, he produced a famous diagram showing how vision was mathematically possible; it is still shown in many art and visual culture classes today.

He showed light entering the eye as a set of geometric lines. He solved the question of how large objects can be seen by showing that the rays are refracted by the eye's lens and converge on the retina at the back of the eye. However, this is not seeing. The image produced on the retina was interpreted by what Descartes called the sense of judgement. The drawing represents judgement as an elderly judge, assessing what there is to be seen and coming to a decision about it. Vision was understood as a courtroom, in which the eye presents evidence for the judge to decide. (There was no jury, as in the French courts of the time.) Descartes' breakthrough not only helped us for the first time to understand how vision was possible. It also raised the importance of seeing to a new level as the key sense in modern science, which centres around the observed experiment.

In our own time, we are witnessing how neurology, a fast-developing part of biological science, sees the body and mind as integrated systems and people as communal, social beings connected by empathy. The metaphors here are not taken from the courtroom but from computer networks. It is a very different way to see ourselves and to think about seeing. According to this perspective, we learn how to become individuals as part of a wider community. This outcome is the intriguing result of the revolution in studying the brain, which many would consider to be the most individual organ of all, and in particular how humans and other primates see. My point is not that modern neuroscience is the final version of the 'truth' and all other previous understandings have been shown to be wrong (although some neuroscience boosters do come close to saying this). Rather, as we shall see, neuroscience and its ways of visualizing the mind and human thought are becoming the vital visual metaphors of our time. It is our version of the truth, for better or worse.

Visualizing vision

In the late 1990s, the psychologist Daniel Simons and his student Christopher Chabris devised what would become a famous experiment: a video test known as the 'Invisible Gorilla' (Chabris and Simons 2010). Those who participated in the study were asked to watch a video and count the number of times the team wearing white passes a basketball while they play a team wearing black shirts. As this simple action unfolded, a person wearing a gorilla suit walks across the court.

Roughly half of the people watching did not even notice the gorilla. They were concentrating on counting. Simons attributes this to what he calls 'inattentional blindness', the inability to perceive outside information when concentrating on a task. Researchers had been aware of this

Figure 22 – Simons and Chabris, still from 'Invisible gorilla' video (1999)

phenomenon since the 1970s, as had magicians from time immemorial – 'the sleight of the hand deceives the eye' because the magician distracts your attention. But it was the video that made the test so dramatic. You could test yourself and then watch the video again to see how obvious the gorilla then appears. Some people get very upset when they realize their failure.

This experiment built on the research of neuroscientists like Humberto Maturana from the 1970s. Maturana demonstrated that a frog, for example, sees very differently from the way we do. It perceives small, fast-moving objects, like the insects it eats, very clearly, while ignoring large slow-moving things. Birds can perceive ultra-violet light invisible to humans, which allows them to see their own plumage differently than we do. However, even this seeing is not vision.

Maturana stressed that living things change themselves because of their awareness of their interactions with the outside world, not just in the very long run described by evolution, but as a condition of day-to-day existence (Maturana 1980).

That is exactly what has happened in response to new media. When I show the Invisible Gorilla test video to students and others today, nearly everyone sees the gorilla. A population that has grown up with video games and touch screens sees things differently. Simons himself has found that when you show the video to experienced basketball players, the number seeing the gorilla jumps to about 70 percent. Simons carried out a more recent study demonstrating that some people continue not to see the gorilla, based on a small sample of 64 people. Of these only 41 were unaware of his video. In this group, 18 did not see the gorilla, well under 50 percent. My sample group is larger in size, and compiled over several years, although not conducted as a scientific study. Perhaps my counter-sample, drawn as it is from participants in visual culture classes, is just more visually aware.

The capacities of the human body obviously cannot have evolved in such a short space of time. Rather, the change comes in the way we make use of visual information. In the age of industrial work, concentrating on a specific activity and ignoring distractions was highly desirable. From academic research by a student in a library to the adjustment of a machine by a factory worker, attention needed to be focused. Today, we prioritize the ability to keep in touch with multiple channels of information – multi-tasking is the popular term. As I write this book, people are sending me emails and text messages to which they expect prompt replies, regardless of what I am doing. Formerly, we were trained to concentrate on one task, meaning we might not see the gorilla, and mostly, though not exclusively, we did not. Now we are trained to pay attention to distractions and mostly, though not exclusively, we do. Neuroscience has changed the way that vision is understood. However, there's still noticeable room for interpreting that change.

Seeing the brain

Let's begin with how we can now 'see' the brain in action. With the invention of new forms of medical imaging, especially Magnetic Resonance Imaging (MRI) in 1977, it became possible to make 'pictures' of the brain at work. Of course, no light is involved, and no drawing or other representational work is done. The magnetic field created by the machine excites hydrogen atoms in the brain (or whatever body part is being examined). As a result, they emit a radio frequency that is detected by the machine and converted into images. It is possible to imagine a species that could hear those frequencies and detect what is wrong (or not) with the person being scanned. Humans need to see something.

An MRI scan is actually an exercise in the history of media. Magnetism and its relation to electricity was a fascination of nineteenth-century science. The Scottish scientist James Clerk Maxwell demonstrated that light itself is a form of electromagnetism and he calculated its speed. He was also able to take the first partially successful colour photograph in 1861. It was the study of electromagnetism that later led Albert Einstein to his theory of relativity and also made possible the invention of radio, the first mass medium. First used to communicate with ships, radio became a popular format in the 1920s. Now, however, it is our own bodies that are the transmitters. As we cannot interpret these waves unaided, the MRI machine converts them into visual form. These are not images, however, in the sense that they are representations of something seen. The scanned organ remains inside the body, unseen by anyone or anything. Like any other picture produced by a computer, MRI scans are computations, not images. Neuroscientists use a particular process called Functional Magnetic Resonance Imaging (fMRI) that allows them to 'see' where the blood flows in the brain in response to specific stimuli. These 'pictures' have developed a new mapping of the brain's functions, showing which structures in the brain 'light up' when an action is carried out.

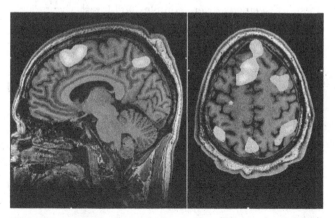

Figure 23 — fMRI scan

These dramatic images seemed to make it clear that the brain has local specialities. The scan in Figure 23 shows a person using their memory and it is clear that certain areas of the brain are in use and others are not.

At the same time that academics, artists and activists were creating visual culture studies, scientists were transforming our understanding of vision itself by using these new techniques. In 1991 Daniel J. Felleman and David C. Van Essen published a now-classic analysis of visual function in primates, based on a study of macaque monkeys, because of their similarity to humans. In their summary, Felleman and Van Essen concluded that they had found

> 25 neocortical areas that are predominantly or exclusively visual in function, plus an additional 7 areas that we regard as visual-association areas on the basis of their extensive visual inputs. A total of 305 connections among these 32 visual and visual-association areas have been reported.[1]

In short, seeing is a very complex and interactive process. It does not, in fact, happen at a single 'place' in the brain, as the first 'lighting up' images had suggested, but all over it in a rapid series of back-and-forth exchanges. Further, this interactivity between the visual zones of the brain and their associated areas happens at a series of ten to fourteen hierarchical levels. That is to say, seeing is not a definitive judgement, as we had once assumed, but a process of mental analysis that goes backwards and forwards between different areas of the brain. It takes a brain to see, not just a pair of eyes.

Felleman and Van Essen created a diagram of vision

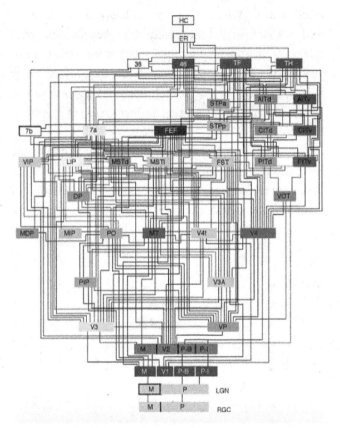

Figure 24 – Felleman and Van Essen, 'Hierarchy of Visual Areas'[2]

in the era of digital computing. In this mapping, the neural pathways for each sense are distinct but are processed in parallel, like a computer. Their understanding of vision shows it as a set of feedback loops. Their map looks quite unlike any earlier model of seeing.

The one point of overlap with Descartes' earlier drawing is that the retina is still included. It's right at the bottom, labelled RGC. What we call vision happens in the set of feedback and parallel processing that takes place between this point and the hippocampus (HC) at the 'top'. Clearly, vision is not just a case of light entering the eyes and being judged, as it was for Descartes. It is a back-and-forth shuffle with twists and turns, creating a vibrant sense of rhythm in the image.

The diagram suggests a visual parallel with one of the great Modernist paintings: Piet Mondrian's 1942 classic *Broadway Boogie Woogie*. Here the Dutch artist adapted his neo-plastic aesthetic to express his sense of jazz-era New

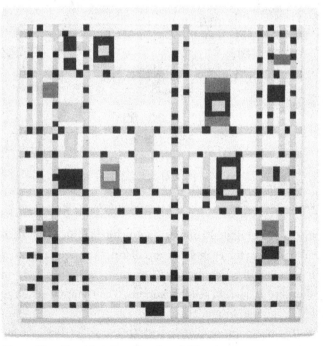

Figure 25 — Mondrian, *Broadway Boogie Woogie*

York. The painting conveys the dynamism of the grid city of Midtown, as well as its unexpected vibrations and drama that make the urban experience anything but regular. Boogie woogie was an uptempo piano form of the blues, often accompanied by a spectacular form of jazz dance that centres on a back-and-forth between the dance partners. In the painting, the combination of repeated bass rhythm and staccato dance movements evocatively conveyed the affect of the machine age. If we take the liberty of making a formal comparison between this painting and the 1991 depiction of vision (recognizing that they come from very different contexts), we can see in both cases how vision has gone from being the single decision imagined by Descartes to the dynamic experience of the modern, machine-based city, conveyed by its flickering lights, back-and-forth journeys and infectious music.

In his analysis of this depiction of visual processing, the celebrated neuroscientist V. S. Ramachandran highlights the importance of the feedback between stages:

> Note especially that there are at least as many fibers (actually many more!) coming back from each stage of processing to an earlier stage as there are fibers going forward from each area into the next area higher up the hierarchy. The classical notion of vision as a stage-by-stage sequential analysis of the image, with increasing sophistication as you go along, is demolished by the existence of so much feedback.[3]

It is not just up to the judge. To extend my metaphor, it's the interplay between city residents on the street, the way the dancer responds to the piano, the feeling of being part of something. The information goes back and forth from one level to the next, filling in pieces as it goes. So, it is (for now) settled: seeing is something we do, rather than something that just happens naturally. More precisely, we need to set aside the persistent notion that an image gets relayed from the retina to the brain – the retina is itself part of the brain where, in Ramachandran's words, 'the rays of light are converted into neural impulses'.[4] From that point, information is distributed and processed in a series of parallel steps that continually reinforce the other layers. What we used to call an image is now known to be a computation, even in the brain.

In the current model, we don't even 'look' to see. In Descartes' version, the judgement presided over seeing, making it seem a very deliberate process. In fact, there are three kinds of unconscious eye movement involved in seeing. Convergence movements direct both eyes to the same place. Pursuit movements track moving objects. Within these overall ways of seeing, research has shown that the building block of close seeing is the 'saccade', a spontaneous scanning by the eyes that move from one point to the next. Saccades are very rapid. Saccades can be set off voluntarily, as when we direct our eyes at a painting. Or they can be involuntary, in response to a moving object, a noise or any other unexpected event. So the old idea of a single gaze or look, fixing people or objects under its stare, has to be modified. Our eyes are always busy, boogieing back and forth. The resulting mental 'image' that we 'see' remains stable because the brain computes it in that way.

The saccades produce data points that allow us to make

calculations, such as how to pick up a coffee cup (for example). We should think of looking as a whole as a form of doing, or a performance, as we called it in Chapter 1. We make a world in which the way we look makes sense and enables the actions we want to perform. And it is also a form of computing because we use that model to calculate how to be active in that world. So if we're trying to count basketball passes, we can easily ignore the gorilla in the famous video described above because it has nothing to do with our counting action. If we are used to observing passes, though, we have more mental space to notice the gorilla. When the actions that we are trying to perform fail, we may simply have failed to throw an apple core into the bin. Or we may be totally disoriented by a hitherto unknown experience, such as an accident or disaster. We continuously rework these systems, absorb information, and change the way we perceive in order to account for it. In playing games that require hand-eye co-ordination, players are familiar with the experience of 'getting their eye in', meaning that after the first few efforts, it seems to become easier to hit the ball. Our minds and bodies are continuously interacting, forming one system.

For all the discoveries of recent decades, the place of visual perception has nonetheless become less precise. Vision now seems more akin to the puzzling one does in front of a complex painting like *Las Meninas* (Chapter 1) – the moving back and forth to gain different effects, the homing in on certain details, the changing affect that often returns to the beginning point – than it does to the instant affirmation of photography. Eighteenth-century art theorists once proposed a theory of *papillotage*, meaning 'flickering' or

Figure 26 — Boucher, *Diana Leaving Her Bath*

'blinking' vision. Painters like François Boucher aimed to create this sense of moving surface that would later come to influence the Impressionists. It accepted the constructed nature of vision. A 'blinking' vision is aware of its effort to see, of the difference between it and what it is trying to see, and even the eyelid that comes in-between the two.

In his small painting *Diana Leaving Her Bath* (1742), Boucher sets off the sensuousness of the goddess of the hunt's naked body with a dramatically vibrant background of golds and greens, representing the natural world behind her. The painting creates a sense of tactility to heighten its allure. By contrast, neo-classical painting, which came into fashion in the late eighteenth century at the time of the French Revolution, stressed the primacy of drawing, using hard edges for figures and objects. If we glance at Jacques-Louis

Figure 27 — David, *Antoine-Laurent Lavoisier and His Wife*

David's much-larger portrait of *Antoine-Laurent Lavoisier and His Wife* (1788), the difference is at once apparent.

There is certainly sexual intrigue here. But the crisp precision of David's lines and almost hallucinatory clarity of the figures and their outlines could not be more distinct from his predecessor Boucher. It was this approach that would be imported into photography as 'correct' focus. This realism wanted to efface all flickering and insist that what we see is what is there and nothing else. Early photographers, like the British artist Julia Margaret Cameron, disputed this, demanding to know in a letter from 1864, 'What is focus?' (Mavor 1999). Do we concentrate on producing sharp lines, or insight into the content of the photograph? In her portrait of Thomas Carlyle, for example, Cameron produced an image of the historian as mystic that is arguably closer to Carlyle's sense of himself as a 'seer' (meaning both a prophet and a person who sees) than a sharp-edged image would have been.

The point is not that earlier periods correctly anticipated neuroscience or that the current ideas of the neuroscientists must be accepted as unquestioned truth. Rather, what these works of art reveal is that the current research fits into a well-established line of thought about vision, even if it is based on very different evidence and in a very different context. Indeed, it is noticeable that people today often put more trust in a less-than-perfect photograph or video that takes an effort to decipher than they do into a professionally finished work, because they suspect that the latter will have been manipulated.

W. J. T. Mitchell, often considered to be the founder of visual culture studies, was prompted by the rise of digital visual culture to propose 'there are no visual media' (2005). By this apparently paradoxical statement, Mitchell meant that all media involve every sense and so it is inaccurate to describe a painting, which is made by an artist touching a canvas with brushes, as being nothing but visual. Now we

Figure 28 — Cameron, *Thomas Carlyle*

can reinforce that interpretation with the understanding that perception is not a single action but a process carefully assembled within the brain. That work centres on what neuroscientists call 'body maps', our sense of where and who we are. Thus we always know what posture our body is in, even with our eyes closed. Think of how you brush off an insect that lands on you. You have to be able to co-ordinate a sense of yourself, which is now called 'proprioception', a sense of the relative positioning of all the different parts of your body, with the perception of an insect on your skin. Then you use a hand to swipe at the bug at sufficient speed to prevent it biting you but without inflicting excessive pain on yourself.

These maps do not always accord with the objective state of our bodies. To take extreme examples, a person withdrawing from addictive drugs may feel insects crawling all over their skin and swipe at them constantly to no effect. A person who has suffered amputation often feels a 'phantom' limb where the actual member used to be. They know perfectly well it's not there, but it hurts, itches and otherwise calls attention to itself. By the same token, an anorexic person may look at their image and see themselves as obese, rather than the very thin body that others would see. In this case, the anorexic has long been assumed to be distorting their own image, which is part of the clinical diagnosis. However, some 'pro-ana' (pro-anorexia) websites and online communities share images of very thin people, including celebrities and models, as 'thinspiration'. So some people with anorexia (as the medical profession sees it) are as aware of their body map as everyone else. They just draw radically different conclusions from it (and, to be clear, not ones that I would personally support).

Some striking experiments have shown that these maps can be relearned. The neurologist Ramachandran performed a famous therapy with phantom-limb patients by rigging up a mirrored box, so that the surviving hand (for example) was reflected by mirrors in the place of the amputated one. Using this simple visual by-pass, people have been able to redraw their body maps so that phantom limbs can be 'moved' out of awkward positions, itches scratched and so on. It seems that all the redundancy in the visualizing system allows for the possibility of such relearning. Stroke victims and even people suffering from chronic pain have been able to benefit from this therapy. The visual information seems to 'overwrite' the other information available. There is more processing space in the brain devoted to vision than all the other senses combined, which might account for why this illusion is so irresistible.

In less dramatic cases, many optical illusions continue to 'work' even if you know that they are not real. Think of the philosopher Wittgenstein's duck-rabbit, a drawing that can look either like a rabbit or a duck, or both, depending on how you see it.

Figure 29 — Wittgenstein, 'Rabbit and Duck'

We might even see it as a collection of lines and shading. By extension, you might take an active choice to see only in one way, like the person with anorexia who sees their thinness as affirmative. Or you can succumb to the illusion and 'move' the amputated limb that no longer exits.

It turns out, according to recent research, that this is because there are two 'streams' of brain activity: one for perception and one for action (Nassi and Callaway 2009). Vision is a plural noun, it appears. One stream (perception) recognizes a friend. The other (action) reaches their hand and shakes it. There are now thought to be over 80 physical locations in the brain that process vision, connected by at least twelve parallel processing pathways. Despite all this input, key 'attributes such as motion, shape and colour must be computed from these sensory cues'. That is to say, we don't 'see' colour, shape or speed so much as work out what they must be. The brain is not a camera. It's a sketch pad. Intriguingly, the medieval theologian St Thomas Aquinas also held that vision was not relevant to determining where an object was or what colour it was, albeit for very different reasons. I take it from this that people have always known in some way that vision is constantly being learned and relearned.

The mirror and the community

One of the most intriguing of all the insights from the new research is that we do indeed learn mostly from each other, rather than by ourselves, and that our brains are specifically designed for that purpose. Sense experience is not individual but common. This was the surprising result of a set of experiments on monkeys by Italian scientists. The Italian group was working on a project to analyse what nerve impulses were produced by a monkey reaching for a peanut. By chance, they discovered that the monkeys watching had exactly the same neurological response as the one actually grabbing the treat. The same part of the brain 'lit up' in the watching monkeys as in those of the monkey doing the action. The result has been reproduced many times in many different contexts and it has led to some radical conclusions.

Neuroscientists have now discovered that we have 'mirror neurons' whose function is to respond to others. Ramachandran calls them '"Gandhi neurons" because they blur the boundary between the self and others – not just metaphorically, but quite literally, since the neuron can't tell the difference'.[5] The quality of empathy is, in the current metaphor, hard-wired. Vittorio Gallese, an Italian neuroscientist, has explored the implications of his experiment with the monkeys described above (2003). For Gallese, the brain is in effect a shared space with a 'we' that is not a crowd of individuals but a collective formation. It is from this formation that individuals emerge, meaning that we move from the social to the individual. By this he means that we don't generalize what we learn on our own, but apply the general to the specific, because we all form a 'theory of mind'. Such a theory, well known to philosophers, is vital to human interaction because without it, no one could begin to imagine how others might act. Like all visual culture, this insight is rooted in the everyday. If, in a café, I see a person reach for a coffee cup, I believe she is going to drink because that conforms to my theory. Of course, she may throw the coffee over her spouse but that would be an exception to which my theory would then have to adapt. Rather than being a distraction from reality, the imagination is key to our very understanding of how we exist in the world.

In short, mirror neurons do not only allow me to see the world from my point of view but to visualize it from the point of view of others. As Ramachandran points out, humans are exceptional because we develop so slowly, learning everything from basic motor skills to language *after* birth, unlike most animals: 'Obviously we must gain some very large advantage from this costly, not to mention risky upfront investment and we do: It's called culture'.[6] So it would not be unreasonable to say that the collective theory of seeing we develop would be something that we can call 'visual culture'. Certainly this culture varies from place to place and time to time. As part of the interfaced sensory apparatus of the brain, visual culture is not only visual (in the common-sense meaning of the term) but is also concerned with the entire body map. The final implication is that humans have developed in the dramatically rapid way that we have, not only by Darwin's natural selection but also through cultural development.

All of the experience we call 'vision' is multiply processed, multiply analysed and subject to constant feedback from other areas of the body, rather than being a single, independent 'sense'. Researchers now want to add many new senses, like the proprioception described above, to the classical five senses. For example, we feel hunger and thirst as strong needs without having to think about them and without external stimulation. But that is not to say, as some sceptical philosophers have done, that we all see differently or that we can make no use of the sensing that we do. Because seeing is a specific instance of our collective theory of mind, vision is a commons, meaning a shared resource that we can nonetheless make use of in ways that also suit our individual needs. And this is what visual culture seeks to debate, explore and explain.

The World of War

If in theory we all share a visual commons, in practice those in political power have always claimed to be able to see differently. As we saw in Chapter 1, kings in particular often claimed to be more than simply human. These unusual abilities were demonstrated when kings led their countries on the battlefield in war. We still use 'leadership' and 'vision' as meaning more or less the same thing from this history of battle. Battle was not without risk, to be sure. To take the case of England, from the Anglo-Saxon Harold, who died at Hastings in 1066, to Richard III, killed at Bosworth Field in 1485, kings (or royals) sometimes paid for this glory with their lives. But the last English king to venture into battle was George II, who participated in the battle of Dettingen during the War of Austrian Succession in 1743.

By the time of the Napoleonic wars (1799–1815), kings and other national leaders were no longer appearing on European battlefields. Modern war spread out over wide expanses of territory. It could not be seen from one place. Kings and other leaders stayed at home and delegated war to their generals, who themselves directed the combat from far behind the engagement. War became known as an art in the West, as it long had been in China, requiring a specific new visual skill, which later came to be called 'visualizing'. The task of the general was now to 'visualize' the battlefield as a whole, even though he could not see it. He had to add his imagination, insight and intuition to whatever he and his subordinates could see for themselves. The general visualized the battlefield as if from the air so as to imagine where his own forces were, the situation of the enemy and how the two might interact. This skill was first the attribute of the general and then became a highly specialized set of technologies. The result was a way to see the world from the air. It is not transparent or simple but presents itself as difficult, and open only to a few highly skilled people, interpreting data from machines.

The eighteenth-century Prussian general and military theorist Carl von Clausewitz (1780–1831) defined modern war in his classic *On War*, based on his experiences in the Napoleonic period. It is still taught in military academies today. He emphasized first that war was mostly invisible:

> The great uncertainty of all data in War is a peculiar difficulty, because all action must, to a certain extent, be planned in a mere twilight, which in addition not infrequently – like the effect of a fog or moonshine – gives to things exaggerated dimensions and an unnatural appearance.[1]

It was now the job of the commander to see through that fog, even though he was not on the battlefield. Clausewitz noted that some of the work was done by sight and some by the mind:

that this whole should present itself vividly to the reason, should become a picture, a mentally drawn map, that this picture should be fixed, that the details should never again separate themselves – all that can only be effected by the mental faculty which we call imagination.[2]

His metaphors are striking: war is a picture physically imprinted or 'fixed' on reason itself. The idea is all the more noticeable because it preceded the inventions of photography or film that it seems to rely on, or more accurately, anticipate. The comment suggests that the death and dismemberment on the field of battle were side-effects of the real event, which was in the mind of the commander. The picture that Clausewitz wanted to see became a very specific kind of image: a view from the air. This imaginary view has become represented as the battlefield map – made after an engagement – which shows the deployment of troops, as if from the air, set into a contour map of the area.

Clausewitz saw that the change was in the very nature of war itself. If war was now a clash between two sovereign powers, rather than specific leaders on a battlefield, Clausewitz is often (if slightly inaccurately) quoted as saying: 'War is politics by other means'.[3] For the visualizers, war is not, then, anything in itself: it is a way of accomplishing political change. What truly matters in modern war is the political result. Visualizing was a way to make this political change happen by means of war.

For Clausewitz, the prime exponent of such visualizing was Napoleon, the French general who made himself Emperor in 1805. Napoleon epitomized the equality of his post-revolutionary times in his simple three-cornered hat and greatcoat. For one of the achievements of the French Revolution (1789–99) was the professionalization of its Great Army, in which advancement was now open to all, not just the aristocracy. War had become the business of disciplined professionals.

Napoleon himself, born in Corsica and transformed from a lieutenant to a general and finally to Emperor, embodied this idea. As a general, he was famous for visually deceptive tactics, such as launching an attack in one highly visible area, only to send a much larger force by a concealed route to attack the troops responding to that manoeuvre. These visualized tactics were central to his famous victories, such as the battle of Austerlitz (1805), where a smaller French army decisively defeated the Austrian and Russian armies. For all that, it was often hard to tell who had won a battle on the ground, as large armies roamed over extensive territories and even the 'winner' suffered high casualties. In Stendhal's classic novel, *The Charterhouse of Parma* (1839), the anti-hero Fabrizio believes that Napoleon has won the battle of Waterloo, which was in fact the final defeat of the Emperor in 1815.

Mapping war

Napoleon had extensive maps made of his battles, including drawings done showing the field from exactly his height so he could later assess what he could and could not see at the time and how well his visualization had worked out. For other less-gifted people, two alternatives soon presented themselves. First, try and see from above by means first of balloons and then aircraft. And then make photographs and other technical images to share that view.

If Clausewitz's metaphor of war as a map now seems 'natural', it was not always so clear, at least in Europe. In China, maps had been used in advanced forms for centuries by the time of Clausewitz. As early as the third century CE, Pei Xiu (224–71) criticized earlier maps for their inaccuracies of scale and distance. Pei Xiu developed both a grid system and means to represent topographical elevation. Using six principles of mapmaking, he created a detailed depiction of the empire. Maps were used for military purposes from the Han dynasty (c. 160 CE) onwards, and tributary states were required to provide detailed maps of their territory to signify their submission to imperial power. By contrast, mapping was to all intents and purposes non-existent in Europe as late as 1400, where maps simply listed the names of settlements in the order required to travel from one place to the next. By 1600, however, maps in the modern sense were crucial to all kinds of activity, from warfare to colonial conquest and land ownership. Still, it was only in the eighteenth century that cartography (the art of making maps) was fully integrated into Western military planning and operations.

And the best place to make a map, which visualizes the world as if seen *from* the air, *is* from the air. In 1794, a French general named Jean-Baptiste Jourdan won the battle of Fleurus by using information on enemy troop movements received from a balloon. By the time of the American Civil War (1861–65), such balloons were equipped with telegraph wires so that information could quickly reach the ground.

From these slow beginnings, maps became so central to modern war that by the end of the Civil War, the Army Corps of Engineers was supplying Union forces with 43,000 maps a year.

It was at the astonishing Berlin Conference of 1885 that mapping as a form of war reached a peak in the annals of 'politics by other means'. At this event, the rulers of Europe took out a map of Africa and calmly divided it among themselves. The resulting division of the continent into colonies had disastrous consequences that continue to this day. For the arbitrary borders they created crossed long-standing national frontiers. When the colonial nations formed in Berlin became independent after World War Two, conflict was almost inevitable. The long war in the Congo and Great Lakes region that has persisted since the Rwanda genocide of 1994, at a cost of an estimated 3.5 million lives, is perhaps the most notorious example. The delegates to the Berlin Conference believed that they were claiming empty space, where the inhabitants had no claim to the land because they did not use it. Europeans had actively forgotten their previous highly detailed knowledge of Africa derived from the slave trade. Seventeenth- and eighteenth-century maps of Africa showed major cities, rivers and other political and physical features. By the late nineteenth century, Africa was simply known as the 'dark continent', a place unknown to Europeans.

This forgetting had significant consequences. It enabled the colonial powers to use the legal concept of *terra nullius*, from Roman law, meaning 'empty land' or 'land belonging to no one'. It designated all land that was not being cultivated by European techniques as empty and therefore available to the first (European) taker, providing that they did then attempt to cultivate it. This classification involved a mode of unseeing. Amazonian Indians, for example, do not lay out fields but encourage the spread of fruit trees and nut bushes as their agriculture – but that was invisible to European colonists. And so the European powers believed their division of Africa to be both lawful and beneficial to the inhabitants, for they were bringing what the missionary David Livingstone called 'commerce, Christianity and civilization'. The shorthand expression for all this law and visualized expansion was 'bringing light to darkness'. By the end of the Berlin Conference, King Leopold II of Belgium was sole ruler of the vast Congo region, with all profits from his ventures there going directly to him, not even to the Belgian people; as if he had won a war, rather than had a map redrawn. The Berlin Conference was precisely war by other means, carried out in order to achieve the political results desired.

When the Nigerian-British artist Yinka Shonibare MBE came to depict the event over a century later, he represented the colonists as headless mannequins, as if to ask, 'Where are the heads of state?' The headless leaders cannot claim 'vision' or the ability to visualize, although they have divided the map of Africa which is inlaid into the table. More striking still are the dazzling multi-coloured fabrics being worn by the figures. These are known as Dutch wax prints, often thought of as being typically 'African' fabrics, but are in fact the product of a far-reaching colonial history. The motifs and designs on the cloth were originally created in Indonesia, where they were known as *batik*. When the Dutch

Figure 30 — Shonibare, *The Scramble for Africa*

became the colonial power, they took these designs home and industrialized their production. In the nineteenth century, these patterns became very popular in West Africa, as they are today, where local elements were introduced to the designs. So what are often sold as 'African' fabrics were in fact produced by the interaction of three continents across the imperial era. Shonibare's installation reminds us both of the violence of colonial visualization and its decapitation of local histories and traditions. In his use of wax prints, he shows us that colonization was not a simple divide-and-rule, as the colonizers liked to think, but a pattern of global histories.

The era of colonial expansion culminated in the disaster of the First World War (1914–18). During the so-called 'Great War', the development of airborne cameras dramatically changed the aerial visualizing of war. Now, information was gathered by aircraft that could fly over enemy lines and observe activity with precision. From this point forward, visualizing was a technology at the service of military leadership, rather than leadership being the ability to visualize. Most battlefield images were taken at approximately 10,000 feet (3 km) using cameras with an 8-inch focal length, resulting in a photograph depicting approximately one square mile. Photographs would then be labelled to indicate when and where they were taken and with what equipment, and correlated with existing paper maps. By 1918, it was possible to take a photograph from 18,000 feet (5.5 km) which, when enlarged, could reveal a footprint in the mud. Human visualizing of war was now redundant.

These photographs were then used to create battlefield maps. In the inter-war period (1918–39), the US Army developed and expanded its doctrine: 'Map as you move'. This need generated a shift from paper maps to what were called 'photomaps', indicating the transition from human ground-based assessment to air-based photographic visualizing. Paper mapping could cover about 100 miles (160 km) a day of new territory. When the German Sixth Army attacked France in 1940, it was exceptionally able to cover over 750 miles (1,200 km) a day, using carefully prepared paper maps, using all available sources from aerial photography to pre-war tourist guides. Because such speedy advance was considered impossible, the German attack was totally unexpected. Responding to the new speed of war, photographic battle maps were created using new five-lens cameras that provided 9-inch negatives. Used in combination with multiplex

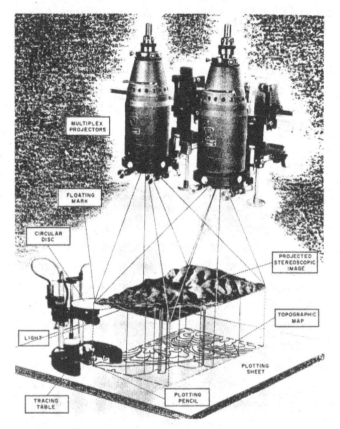

Figure 31 — 'Multiplex Set'

projectors, these photographs could even generate 3-D topographic maps (see Figure 31).

Even on the battlefield, mobile vans were able to turn out thousands of 17×19-inch prints an hour, twice the size of the standard 10×8-inch artistic print. Aerial photographs taken from 20,000 or 30,000 feet (6–10 km) using the new cameras were astonishingly detailed.

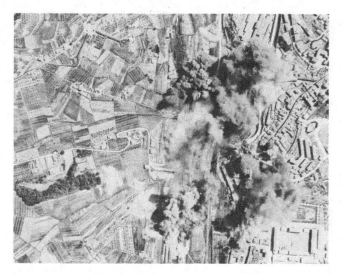

Figure 32 – Aerial photograph during a bombing raid

The photograph of US bombing in Italy during the Second World War, for example, looks like later Abstract Expressionist painting, though of course it is all too real. From this point, all war became air war, meaning that the outcome of war was usually determined by control of the air.

More importantly from our point of view, Clausewitz's concept of war as politics by other means now made visual images the key to the political issues at stake. Nowhere was this more dramatic than in the Cuban Missile Crisis of 1962. One year after the outgoing President Dwight D. Eisenhower had warned of an emerging 'military-industrial complex', global visual technology brought the world to the brink of nuclear war. In 1956, the CIA introduced into service the top-secret U2 spy plane. Flying at an extraordinary height of 70,000 feet (21 km), in the upper reaches of the atmosphere,

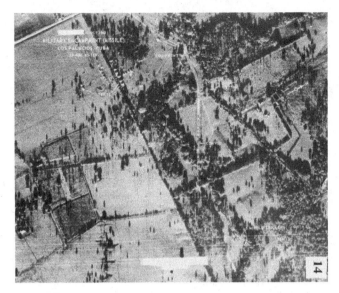

Figure 33 — U2 aerial photograph of Cuba

the U2 was equipped with a high-resolution camera so sensitive that the pilot had to switch off the engines and glide while taking pictures. As a result, it could detect from the edge of space objects two and a half feet long (76 cm). On 14 October 1962, a plane flown by Major Richard S. Heyser took pictures of Cuba that showed preparations for a nuclear missile site.

Even though the picture was taken from three times the height of the Second World War photographs, it is far more detailed and close up. These carefully annotated photographs have become the classic example of what it means to have overwhelming visual evidence. The Soviet Union did not deny that the site was to be used for missiles and the world entered a tense two-week stand-off. Despite the

United States' demand that the Soviets withdraw all materials, ships carrying weapons were heading for the US naval blockade of Cuba. Radio broadcasts reported live as a confrontation appeared inevitable. In fact, secret negotiations were underway to arrange a deal, whereby the Soviets withdrew their missiles in exchange for the US withdrawing some of theirs from Turkey. At what seemed to listeners to be the last minute, the Soviet ships turned around. The moment encapsulated the Cold War. On the one hand, the most advanced visual technology had transformed the nature of conflict. On the other, leaders were still engaging in their own strategies with disregard for the opinions or feelings of their population. The global perception was, however, in the words of US Secretary of State Dean Rusk, that the two superpowers had gone 'eyeball to eyeball and the other fellow blinked'. Kennedy made major political gains by using the aerial surveillance in avoiding war, appearing like a leader and removing missiles from Cuba.

The war of images

With the collapse of the Soviet Union in 1991, the Cold War came to an end. The United States found itself with an overwhelming military dominance, matched by its capacity in visual technology. Visual culture scholar W. J. T. Mitchell has nonetheless named the period since 9/11, a decade after the end of the Cold War, as 'the war of images' (2011). If visualizing was the task of the nineteenth-century general, today images are frequently used as weapons in the war of ideas. As Clausewitz would have understood, the primary task of the images here is to accomplish political goals because they cannot do physical damage like conventional weapons. But they can nonetheless lead to real suffering very quickly.

It is important to note that this shift to a war of images occurred just as the transition to a young, urban, networked global society began to take hold. Although he had fired on revolutionaries in Paris as a young officer, Napoleon always said that his visualizing tactics did not work in cities or mountains. These are most often the terrain of today's insurgencies. Furthermore, most people now live in cities. The era of the battlefield is over, but it has been replaced by the anxiety that anywhere on the planet might become the site of insurgency or terrorism at any time. And in the digital age, the war of images changes the traditional balance of power. A remarkable example in 2014 came with the use of videotaped executions by the group known as Islamic State or ISIS. These horrifying scenes prompted President Obama to abandon his past insistence that ISIS were a 'junior varsity' team not worthy of US concern and make a declaration of extended war against them. Any one person's suffering should be enough to engage our efforts to prevent it of course. For these videos to mobilize the largest army in the world is nonetheless striking.

A useful way to note the transformation is to compare General Colin Powell's presentation to the United Nations in 2003 of his case for war against Iraq with the Cuban Missile Crisis. Powell claimed Iraq's ownership of weapons of mass destruction to be based on 'solid sources', above all a set of carefully highlighted photographs. The 1962 photographs from Cuba had simply shown missile trailers. Powell

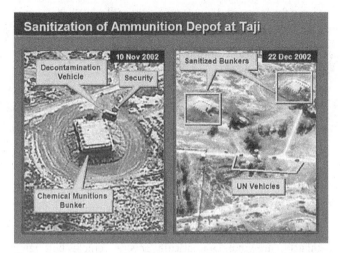

Figure 34 – Slide from Powell's presentation to the United Nations

wanted to make a more complex case. He asserted that his photographs not only showed that Iraq had manufactured chemical weapons but that they had gone to considerable lengths to conceal them. In 1962, the photograph stood by itself. In 2003, two images were incorporated into a highly annotated PowerPoint slide.

Whereas in 1962, the captions were discreet and simply descriptive, the 2003 annotations dominate the photographs. This was perhaps the first political use of Microsoft's PowerPoint software, which is designed to allow simple comparisons of this kind.

Powell distinguished ordinary seeing from specialized visualizing, telling the UN that 'the photos that I am about to show you are sometimes hard for the average person to interpret, hard for me. The painstaking work of photo analysis takes experts with years and years of experience, poring for hours and hours over light tables'. Whereas once it would have been the general interpreting the visualization of the battlefield, he now has specialized technicians to do it for him. These readings are too difficult for ordinary people, even generals. So the bright yellow tags indicate the argument for us. The photograph on the left tags a nondescript structure as a 'Chemical Munitions Bunker'. However, the implication was that the UN should deduce its function from the security installation and the decontamination vehicle, present in case of accidents. In the photograph on the right, two buildings are highlighted as 'sanitized bunkers', meaning that the decontamination vehicle and security installation are gone. Dramatically, several vehicles belonging to the United Nations are arriving, implying that the deception has fooled the weapons inspectors sent to Iraq. Why was all this done? Powell was quite clear: 'There's only one answer to the why: to deceive, to hide, to keep from the inspectors.' In 1962, a clearly seen trailer was what it was. In 2003, Powell's photographs claimed to show what we could not in fact see.

Jorn Siljeholm, a Norwegian inspector, told the Associated Press on 19 March 2003 that what was labelled as a decontamination vehicle was actually a water truck. It was too late. Later that same day, Coalition forces bombed Baghdad, beginning the Second Gulf War. Although millions demonstrated against the war, then-President George W. Bush acidly remarked that he did not pay attention to 'focus groups'. The visualized evidence outweighed any consideration of public opinion because it was the result of expert analysis. Only later did the world discover that there were no weapons of mass destruction in Iraq. The picture was falsely annotated, although whether in good faith or not still remains unclear.

The war of images was reinforced by the new logic of Bush's Global War on Terror, according to which every space on earth was 'either with us or against us'. The task of visualizing the battlefield from the air now extended across the world's surface. During the Cold War, the world had been divided into two zones, Soviet and Western. Conflict took place at the borders between these zones or in contested areas, like Vietnam. The United States and Soviet Union themselves would only have become war zones if the situation had broken down into a 'hot' war, as was threatened during the Cuban Missile Crisis. At the time, this was known as 'containment'. After 9/11, the entire world became a place for possible conflict, although this would mean counter-insurgency or regional war, rather than classic battles, let alone global nuclear war.

Unlike the Cold War with its 'balance of power', today's hostilities are not an even conflict. Like all insurgency, the war of images is an asymmetric war, meaning that the United States and its allies have an overwhelming superiority of force but can still be drawn into exchanges which cannot be simply 'won' in the way a formal battle might be. These wars are also asymmetric in that partisans of each side feel themselves to be overwhelmingly in the right and will not concede that the other side even has a case.

The war of images has similarly unfolded in asymmetric ways. The 9/11 plotters, for example, intended the attacks above all to create spectacular media images, with the first plane drawing in countless media in time for the second attack. When the United States opened a detention centre at Guantánamo Bay, Cuba, it released photographs of detainees dressed in orange jumpsuits, wearing sensory-deprivation headphones and goggles. The photographs were 'answered' by appalling videotaped executions of Western hostages forced to wear similar jumpsuits by self-proclaimed jihadis. The Internet, first created by the US military to enable the exchange of messages in the event of a nuclear war (Abbate 1999), is now the asymmetric medium by which these gruesome films are distributed. There is a push to see the obscene here, in the literal sense. Obscenity is what is kept off the scene, or off stage. It used to mean the violence of Greek tragedy that was not seen by the audience but only reported to it verbally. In the image war, it is important to have violence shown in order to claim a victory. When the reporter James Foley was executed by Islamic State (or ISIS) in 2014, the Metropolitan Police in London asserted that 'viewing, downloading or disseminating extremist material within the UK may constitute an offence under Terrorism legislation'. When pressed by journalists, government sources could not say which laws were in fact being broken. The concern was nonetheless clear: young British people are being drawn into 'jihad' by such media and the police want to break the connection.

In the effort to forestall such 'blowback', as the CIA calls such rebounds, the high point of the era of image war was the Shock and Awe bombing that began the invasion of Iraq. In 1996, a paper presented to the National Defense University argued that the US should now seek 'to affect the will, perception, and understanding of the adversary to fight or respond to our strategic policy ends through impos- ing a regime of Shock and Awe'.[4] The mass bombing of Iraq in March 2003 was the first real effort at using the strategy, which was designed to reduce the length of war and result- ing civilian casualties by demoralizing both the enemy and any domestic opponents of the war (Mirzoeff 2005). Just as with the 9/11 attacks, a key aspect of Shock and Awe was that it needed to be seen by those not directly attacked.

The image war in Iraq was dramatically successful at the level of the state and unsuccessful across the Iraqi popula- tion as a whole. Saddam Hussein's government collapsed almost at once. But an unexpected insurgency was being prepared at the same time, which subsequently engaged with asymmetric image warfare. So we saw things we thought we would never see, like burned bodies hanging from a bridge in Fallujah. If you knew where to look, or were willing to do so, there were repeated 'trophy' photographs online, in which soldiers posted unedited digital photographs show- ing dead bodies. There were insurgent videos, claiming to show explosions of Improvised Explosive Devices and other attacks on Coalition forces. These pictures showed appalling violence but quickly became almost banal, through sheer repetition.

The obscenity of the image war went public when the Abu Ghraib photographs of detainees held by the United States in Iraq were revealed in April 2004. That the prison had been built and used by the former dictator Saddam Hussein as a torture site added to the shock. Even today, not all the photographs have been released, but those that were are shocking enough. Prisoners were forced to engage in same-sex sex acts, to appear naked, to crawl on leashes. Prisoners were made to stand in 'stress positions' with their arms extended or hanging by their arms. In later investigations, it was nonetheless determined that, other than the sexualized punishments, most of these actions were deemed to be following 'standard operating procedure' (Gourevitch and Morris 2008). The scandal of the photographs was, nonetheless, the visible sight of pain and suffering in a war in which the Coalition forces did everything they could to erase such evidence. Furthermore, the sudden appearance of these photographs undid the asymmetry of the 'good' war being presented by most Western media at that point, in which the Coalition was ousting a violent dictatorship in order to create democracy. Considered simply as images, ignoring the political context of the image war, the photographs are wildly disturbing, and were among the most famous of the Iraq war. Yet ten years after the scandal, many students I speak with, who were children at the time, are unaware of what these photographs are, and say they have not seen them before.

The rise of the drones

If the Abu Ghraib scandal was shocking, it was also visible evidence that the war of images had not succeeded on the ground. The US-led Coalition had failed to 'shock and awe' the resistance. Whatever the cause, the Coalition was seemingly unable to prevent the situation from worsening, despite the successes of the 'surge' in Iraq (2007–8).

Asymmetric warfare is very hard to win: there are no capitals to conquer, nowhere to hoist a flag. Relatively small acts can make it seem that the conflict is still ongoing, as we have seen in both Iraq and Afghanistan. Once launched, the image wars have proved very hard to contain. Some events have astonishing consequences, while others, of apparently equal moment, disappear. It is sometimes no longer clear how politics can be conducted at all, whether by war or other means. So the decision has been to launch war by other means.

War has gone back into the air – but with a twist. The now ubiquitous Unmanned Aerial Vehicle (UAV) or drone visualizes its operations from above, consistent with the long history of seeing the world as a battlefield from the air. Whereas Shock and Awe hoped to subdue entire populations, the aim now is to see a target and then eliminate it. The goal is to decapitate any resistance by depriving it of leadership. There is no longer a battlefield, only zones of surveillance. Those zones have moved beyond the official conflict areas to all the major areas of government concern that have been designated as 'wars', in the metaphorical sense, such as border security and drugs. The drone literally makes politics into war by other means. Political officials decide whether or not to target specific individuals and even watch the results.

The key moment in the shift to aerial counterinsurgency came on 1 May 2011, when US Special Forces carried out the targeted assassination of Osama bin Laden. The killing of bin Laden marked an unannounced but clear shift in the global counterinsurgency to a policy of targeted assassination, wherever the subjects may happen to be.

Figure 35 — Obama watching the bin Laden raid

The official visual record of the event showed President Obama and other leading officials watching what appeared to be live video feed of the mission. While the leaders saw what happened, Obama refused to release a photograph of bin Laden's killing and his body was disposed of at sea before the news was even announced. In refusing to release bin Laden's photograph, Obama gave notice, in effect, that the asymmetric 'war of images' was over. The 'victory' attained from a photograph or video proving bin Laden to be dead would be more than offset by its use as propaganda for al-Qaeda.

This extraordinary execution set the way for the drone war that has followed. Drones are, in effect, flying video cameras that usually have missiles attached. The Predator drone (see Figure 36) carries a Hellfire missile, which has a 20-pound (9-kg) high-explosive warhead. The operators sit in trailers at air force bases thousands of miles away. They

Figure 36 — MQ-1 predator drone

fly the machines using precisely the kind of joystick a gamer would use. The drones can go anywhere but the video feed that they send back is far less precise and detailed than the spy-plane photographs. Lacking depth, the blurry flat images become hard to interpret. The operators call the people they kill 'bug splats', meaning the way that an insect's body splatters on the windshield of a car. Analysts in leaked transcripts debate whether something is a weapon or not. A photograph from a U2 is clear to see, but by the time it has been processed, the person with the weapon, if that's what it was, would have gone. Now there is the option to shoot before the presumed insurgent's weapon has even been used.

In May 2012, the White House revealed to the *New York Times* that President Obama personally ruled once a week on the intelligence agency 'kill or capture' lists of names. If your name is on this list, the agencies will attempt to kill you by drone strike or other means and claim legal protection for doing so. Even American citizens have been so targeted. This is one of the classic exercises of sovereign power. One of the oldest rights of kings has been *droit de glaive*, literally the right of the sword, meaning the power to decide who lives and who dies. Jury trial and other apparatus have much diminished that power over time, until digital technology suddenly restored it. While a commander-in-chief is required to have the authorization of Congress to go to war (at least in theory), these killings are decided by the executive branch alone. Here politics is again war by other means. The goal is no longer to win the war, but to make sufficient political gains, especially at home, to justify the action. Seen in this way, it is perhaps less surprising that the current means of visualized war are missiles fired from drones, controlled from home territory, based on sovereign decisions also taken remotely, at home. The long history of removing the general from the battlefield, while enhancing the possibilities for visualizing the conflict, has reached a (literally) new high point.

Drones are now becoming global. In early 2013, the US established a drone base in Niger, one of the world's poorest countries, as concerns grew about Islamic radicalism in Mali, Chad, Sudan and elsewhere. The military are developing a new class of Micro Aerial Vehicles (MAV) that are only 15 inches across or less.

Larger MAVs, like the unit in Figure 37, are designed for longer-range reconnaissance. Thousands of such MAV units could be deployed for the cost of one UAV today. Drones are now proliferating, extending the militarized way of seeing the battlefield to other arenas. Their most effective use

Figure 37 — Black hornet nano UAV

is as surveillance platforms. In 2013, showing that the whole world really is becoming a form of visual battlefield, Amazon claimed that their packages would soon be delivered by drone. The Federal Aviation Authority intends to integrate such drones into federally regulated airspace in the United States as early as 2015. Police forces worldwide are eager to use small drones, just as activist groups hope to use them to monitor police. Drones have spread from active areas of US counterinsurgency in Afghanistan, Iraq and Yemen to undertaking aerial surveillance of the US–Mexico border, of drug cartels in Central America and most recently of insurgent activity in North and West Africa.

However, the results are not always so clear. A video released by Wikileaks in 2010 showed a local Reuters journalist being killed from a helicopter in 2007 after the misrecognition of his camera as a weapon. Although the drone pilots don't make the decision whether to fire, it seems that video game shoot-first tactics predominate. According to the famously fact-checked Index in *Harper's Magazine*, only two of the 607 people killed by drones in Pakistan in 2012 were on the 'most wanted' list. Only two percent were Taliban or al-Qaeda leaders.[5] It's not surprising, then, that governments in Pakistan and Afghanistan are increasingly refusing to co-operate with the US military, given this level of civilian casualties. In the absence of official figures, it is hard to be sure how many casualties there have been. Columbia Law School's report estimated that between 456 and 661 people were killed by drones in Pakistan alone in 2011, of whom between 72 and 155 were civilians (Columbia Law School 2012).

These are small numbers by the standards of industrialized war and pale by comparison with the millions of deaths in the world wars or post-colonial conflicts. There is a certain unique horror to the very precision of the process nonetheless. To be seen by a drone is to be under potential sentence of death. Journalist Steve Coll quoted Malik Jalal, a tribal leader in North Waziristan, in 2014 to the effect that 'Drones may kill relatively few, but they terrify many more'.[6] Further, as the British artist James Bridle has put it, the drone 'embodies so many of the qualities of the network. Sight at a distance, action at a distance, and it's invisible'.[7] In other words, it's very easy to imagine being targeted by a drone because we are already living in the network that makes it possible. More than that, it feels as if the 'dead' network is coming alive and watching us, in a digital version of the zombie theme that is so popular in today's film and television. As citizens worldwide feel themselves to be under more and more surveillance, it can seem that policing by drone is the future for all of us.

There is already a widespread amateur and commercial sector using small drones. Built from approximately the same digital components as an iPhone – a processor, battery, GPS and a camera – these drones are envisaged as doing the same tasks that light aircraft and helicopters do now, except at far lower cost. That might mean surveillance on power lines or crops, looking for lost hikers, or spraying pesticides. Sales of prototypes are brisk. It's as if our phones have come to life, taken to the air and started watching us.

The drone epitomizes the new moment in global visual culture. It creates a view from the air via endless low-quality images that are hard to analyse, even as they are connected to lethal missiles. It is a networked device, deployed globally. With the emergence of the new micro-drones, the future of the UAV seems to be in the cities where traditional armies are less able to work. In order to understand the emerging world of the drone, then, we have to go to its territory: the network and the city.

The World on Screen

In 1895, two brothers arranged for a new distraction in the frenzied modern city of Paris. Auguste and Louis Lumière rented a billiard hall downstairs in the Grand Café to show their version of the new moving pictures. All kinds of entertainments had tried to create the illusion of moving images. One Louis Aimé Le Prince, a French photographer, was the first to think of cutting up photographs, mounting them on a strip and projecting light through them in 1888. Thomas Edison had patented his Kinetoscope in 1891 and held public demonstrations from 1894. The human eye retains an image for an instant after it perceives it, a phenomenon known as the persistence of vision. The result is that if more than 12 frames are shown in a second, an illusion of movement occurs. Edison showed 48 frames per second, making his device noisy. The Lumières opted for 16–20 frames per second and adapted a sewing-machine mechanism to make the transitions smoother. We now see these screenings as one of the beginnings of modern cinema.

Since 1895, we have watched the world as moving pictures on screen. The world we see has in turn been shaped and ordered by the way we see it, from film to television and today's digital networks. The difference is that whereas we once had to go somewhere specific to watch a screen, the screens now go everywhere we do. This chapter considers railways and digital media as two material networks that produced different ways to see the world. The railway network that enabled the Industrial Revolution from 1840 onwards interfaced with cinema to create one form of visual world. Today, the distributed networks created by the Internet are producing another world that we see on small pixelated screens. While we know how the development of railways changed the world, we are only just beginning to experience and understand the impact of portable screen culture.

Scene from the train

In many accounts, the Lumières' film *The Arrival of a Train at La Ciotat* is said to be their very first. In fact, they did not film a train until 1896 and the 50-second film that is now familiar to us was shot in 1897.

What seems to have in fact been their first effort was shot on 28 December 1895. Entitled *Workers Leaving the Factory*, it showed exactly what the title suggests. The women we see here were working in the successful factory, owned by the Lumières, which made photographic plates, giving the brothers the time and resources to experiment with other projects. If we take these two celebrated films together, we can see how the industrial time and space created by factory work and the railway produced a way to see a world of its own: moving pictures.

In these brief films, shot in one continuous take, a moment of action unfolds. Workers, many of them women,

Figure 38 — Still from Lumière brothers' *Arrival of the Train in the Station, La Ciotat*

Figure 39 — Still from Lumière brothers' *Workers Leaving the Factory*

leave the factory at the end of their allotted work period and pass by the camera. A train arrives in a station. According to a popular tale, the audience at the first screening were so startled by the sight of a train moving towards them that they fled the room. Film historians have shown this to be false, not least because the film of the train was not shown (Loiperdinger and Elzer 2004). It is striking that the subjects of these first ever moving pictures studiously ignore the fact that they are being filmed. No doubt the Lumières instructed their employees to do so, and arranged for the visual surprises that appear in the crowd of women: first, a man on a bicycle, and then a horse and carriage. The *Arrival of a Train at La Ciotat* was similarly choreographed, with several family members appearing in the film.

The apparently simple fact of film shoots at a factory and a train station marks the convergence of several key forces of modernity. Writers of the period had compared the decade of railway expansion of 1847–57 with the era of the European encounters with the Americas after Columbus. That is to say, the railways were world-creating, just as today's combination of financial globalization and networked computers has created the 24/7 constantly-updating-and-refreshed world in its own image. The railway created a new world economy which produced its own time and space; this has been seen as leading to the invention of moving images. The first moving images were, after all, those that people saw from the windows of trains (Schivelbusch 1987).

As the age of rail was beginning, political philosopher Karl Marx used the metaphor of its technology as a world-view ([1859] 1977). It was no more surprising that a mid-nineteenth-century European would use a railway metaphor than it is to hear someone comparing the mind to a computer today. In what became very influential terms, Marx claimed that human society and consciousness are what he called the superstructure, resting on the economic infrastructure of factories, mines and other forms of production. These were terms taken directly from the railway. Infrastructure meant tracks and associated systems, while the superstructure was the train. In short, for Marx, the human mind was a train running on a set of economic tracks.

The railway changed the way people lived, creating its own time and space. The modern time zones still in use today were first devised so that accurate railway timetables could be created. Until then, local time was specific to each place. British railways took London time as their standard. The resulting Greenwich Mean Time was adopted by most national clocks by 1855, although the legal profession continued to use local time until 1880. A similar pattern was followed in the United States. In 1883 US railroad companies created standardized time zones, only legalized by Congress in 1918. Whereas time had been highly local, it now became uniform for wide areas and then changed abruptly at arbitrary points. Another way to describe this would be to say that time before the train was analogue, meaning that it calibrated evenly with each place's relation to the sun. Afterwards, it became digital, meaning that it shifted in arbitrary units of an hour (like the one or the zero in the computer).

Factories made this change real for the new industrial workforce. At the beginning of industrialization, workers would wander out during the work day as they felt inclined, or take naps if they were tired (Thompson 1991). They brought the habits of agricultural life into industrial practice. Soon, though, it came to feel 'natural' that there is a working day and that as much of that day as possible should be devoted to work. Employers and employees battled then and now to extend or shorten the hours of that day. The creation of the railway network enabled people to live outside city centres and travel there for work. By 1910, a third of all French people were season-ticket holders on the train, travelling in and out of urban centres on a daily basis. A century later, France was still the leading European nation in terms of rail transport with 88 billion passenger-kilometres of rail journeys a year.

Both the scenes filmed by the Lumières depicted a tremendous and visible imposition of abstract order onto time and space. The train arrives at a given time, just as the timetable says it should, enabled by the network of tracks across the country. Passengers wait until the train has slowed before opening the doors, while those waiting hang back to allow them off. Our bodies adjust to the rhythm and demands of the machine. In the factory scene, the symbolic swinging open of the gates at the end of the day, followed by an orderly procession of calm workers streaming home, was nothing less than a visual representation of the triumph of work-discipline.

The train and the cinema merged in the form of the tracking shot. In the Lumière brothers' films, the sense of movement came from the train or from the people leaving. The camera itself remained static. Around the time of the First World War, directors began to mount cameras on what is known as a 'dolly', a trolley on wheels. The dolly was and is moved along a set of tracks, directly imitating railway tracks. If the cinematic view is the scene seen from the train, the shot is made from the tracks. Around this time moving pictures became less of a novelty and audiences began to decline. Some operators sat spectators in spaces designed to resemble train carriages and projected their images on the windows. The train became a cinema and a place of modern spectacle.

The avant-garde Soviet film director Dziga Vertov wanted to liberate the camera from the limitations of the human eye and offer a way of seeing that the unaided eye could not achieve. It began with the train:

> I was returning from the railroad station. In my ears, there remained chugs and bursts of steam from a departing train. Somebody cries in laughter, a whistle, the station bell, the clanking locomotive . . . whispers, shouts, farewells. And walking away I thought, I need to find a machine not only to describe but to register, to photograph these sounds . . . To organize the visual world and not the audible world? Is this the answer?[1]

And so Vertov came to imagine the camera as a new form of sensory organization as a whole, affecting more than just sight. This is the meaning of his famous punning declaration in 1919: 'I am eye. I am a mechanical eye. I, a machine show you the world as only I can see it'.[2]

Along with other artists, Vertov took cinema to the Soviet people on the Agit Trains, which brought art to localities outside the urban centres. Films were screened to rural audiences directly from the train. In moments such as this, the train, visual culture and the ideal of the modern interacted to create new ways of seeing the world. Trains made visible worlds: worlds to live in, to work in and to imagine yourself in.

Closed worlds

In the Cold War era (1945–1991), classic studio films like *Brief Encounter* (1945, director David Lean), *3.10 to Yuma* (1957, director Delma Daves) or *Lady on a Train* (director Charles David, 1945) made the train the central place of action. Love and murder now took place on the train. Cinema had moved from being what was seen from the train to being set on the train. These were metaphorical depictions of what historian Paul Edwards has called the 'closed world' imagined by the Cold War, 'within which every event was interpreted as part of a titanic struggle between the superpowers' (1996). Edwards stresses that 'metaphors, technique and fiction' were just as important as weapons systems and computers in constructing the hyperbolic belief that every aspect of global life could be monitored and controlled. Cinema used the train as a key metaphor for closed worlds, and made them believable.

In 1951 director Alfred Hitchcock brought these themes together in his classic *Strangers on a Train*. In the dramatic opening sequence, we see as the camera does, or more exactly, as if we were a camera. The camera tracks two men arriving at a railway station. We see only their shoes, enough to tell us that one is something of a dandy because he has

Figure 40 — Still from Hitchcock's *Strangers on a Train*

two-tone black-and-white shoes, while the other is wearing the then-standard black leather and carrying tennis rackets. Then we cut to a view as if from the train itself.

The camera sees like the train, and the tracks fill the screen. Here the train is acting as the dolly for the camera, so that 'I am a camera' has also become 'I am a train'. The train is a closed world that must follow set tracks. Just then, it changes tracks, carrying our gaze with it, showing that we are not free to look but set on a particular path. Cut. We learn the identities of the two men we saw entering the station, for they have both boarded 'our' train and are sitting opposite each other. The dandy Bruno Antony (Robert Walker) recognizes the tennis player Guy Haines (Farley Granger) and engages him in a conversation about his desire to 'do something'.

Over lunch in his compartment, Bruno presents Guy with his theory of the perfect murder. Two people meet on a train, each wanting a person in their life killed. In their case, Bruno hates his father and Guy needs to escape his wife so he can remarry. So each does the other's murder and there's no way to connect them to the crime. At the exact moment Bruno begins to outline his scheme, Hitchcock alters the camera's viewpoint. Until this point it has been tracking each speaker, cutting back and forth in the standard process known as shot/reverse shot. This is intended to make the viewer feel part of the conversation. The new view allows us to see the whole scene.

Bruno is on the right, as we see it. The blurry line is a telegraph pole passing the window, left in to make the viewer believe that they are really witnessing what is seen on a train. The cut makes us aware that we are watching a film, just as all moving images are descended from the view from the train. What matters now is what happens within the closed world of the train, not what we can see from it. The

Figure 41 — Still from *Strangers on a Train*

closed world became the preferred environment of the 'male gaze', discussed in Chapter 1. The male characters move the plot forward but they are able to do so only within the limited options presented.

Jump to 1967 and another scene on a train in the avant-garde film *La Chinoise*, directed by Jean-Luc Godard. The jump cut makes a leap across time and space. Once considered very daring because the audience might not be able to follow the changes, the jump cut is now standard TV fare in shows like *Law and Order*, announced in their case with the signature 'da-dum' sound. *La Chinoise* is a highly stylized film and is often difficult to watch. In the middle of the film, however, there's a dramatic exchange between Véronique (Anne Wiazemsky), a young Maoist militant, and the activist-philosopher Francis Jeanson, playing himself. Godard posed his actors either side of the train window, which is in view throughout, reminding us again of the moving image as that which is seen from the train by the camera. Inter-titles spell out the French phrase *en train de*, meaning in the middle of something. The pun is intended to show that the train is a space in between, a space between the infrastructure of technology (the tracks), the superstructure of ideas (the carriage space) and the actions that connect them. That space is a visual rhyme for cinema itself. When we are *en train de* we are in the closed world created by the overlap between cinematic and railway networks.

In Godard's scene, Véronique describes her hopes for a Maoist action to close universities and drive students into real-world encounters. At first Jeanson is interested and describes his own project for cultural action. Then he asks

But your action
will lead to nothing

Figure 42 — Still from Godard's *La Chinoise*

Véronique how she plans to do it and she proposes violent action – terrorism, in effect, as Jeanson says.

He contrasts her proposal to the Algerian revolution, in which he was involved, which was supported by the people as a whole rather than a small group. He has the advantage over her, right until the moment when she turns the tables on him by pointing out that she is still a student, and therefore oppressed, and he is not. Jeanson concedes this point, even though he clearly does not think she can succeed. Véronique's insistence on the new privilege of youth and the possibility of transcending the limits of the closed world, despite her naïve approach, gains her the advantage. A year later the student uprising of 1968 seemed retrospectively to vindicate her position.

We see less of trains in Western movies these days than

we once did. While that is in part due to the reduced dependency on trains, especially in the US, it may also be because the cultural significance, and therefore the cinematic resonance, of the train has changed even more. For many passengers, trains are just a suburban connection, rather than the avatar of progressive modernity. In cinema, trains are now haunted by their use in the Holocaust to transport millions to their deaths in extermination camps. As the marathon documentary film *Shoah* (director Claude Lanzmann, 1985) made clear, the Holocaust would have been impossible without the careful co-ordination of train services. His nine-and-a-half hour film used no archival or fictional footage. It shows only the accounts of survivors, witnesses and perpetrators and then-present-day shots of the historic sites. One of the most powerful cinematic moments comes only 43 minutes into the film.

Figure 43 — Still from Lanzmann's *Shoah*

After some witness accounts of the disappearance of Jews from Polish towns and cities, we cut to the scene above. The camera rolls towards what survives of the entrance to the extermination camp at Birkenau. No music or commentary intrudes. We feel the doubling of the camera and the train in a very different way. This is the final culmination of the 'dolly' or tracking shot, where it plunges into the abyss of one of the modern era's worst atrocities. It takes us from a safe space of watching a documentary to an uncanny representation of arriving at Auschwitz-Birkenau, a view that overlaps that of the deported. We can see the weeds growing on the track and the empty surroundings, so the illusion is obvious but nonetheless effective.

In 1985, Poland was still part of the Warsaw Pact, behind what was then known as the Iron Curtain. Relatively few people had visited and the image of the camp was far less familiar than it is today. Just to see Auschwitz was a shock, let alone to arrive there. This connection may not pop into the mind of a train traveller on a busy day but it has become inescapable for film makers after three decades of Holocaust cinema, from *Sophie's Choice* (1982) to *Schindler's List* (1993), *The Reader* (2008), and so many more. In each film, trains play a key role. The train's connection to the violence of modern Europe is more visible now than any concept of it as an icon of progress.

By contrast, the 150 million migrant workers in China, on whom the global digital economy depends for cheap labour and cheap products, make their way to the factories of the special economic areas and back home by train. At Chinese New Year in January 2014, newspaper reports estimated 3.6

billion train trips to make the traditional return home for the holiday were taken by migrant workers; this was (in effect) the largest migration in human history. These workers build most of the computers, phones and tablets on which Westerners airily write about the end of the real and the demise of the train. Given the enormous role of train travel in Asia, it's not surprising that it continues to appear in its cinema. Trains are very advanced in East Asia with the 430-kilometre-an-hour (268 mph) Maglev bullet trains of China and Japan outpacing all European and American equivalents.

In Hong Kong director Wong Kar-wai's stylish and evocative film 2046 (2004), the hi-tech train plays a key role. The film is influenced by both Hitchcock and Godard, absorbing the Western film-noir and avant-garde styles into the Chinese closed world. As if in a Hitchcock film, the leading male character Chow Mo-wan (Tony Leung) often spies on women through convenient peep holes. Meanwhile, the film is visually stylized in the fashion of Godard and follows his habit of sparse, enigmatic dialogue: 'Every passenger who goes to 2046 has the same intention. They want to recapture lost memories because nothing ever changes in 2046. Nobody knows if that's true because nobody's ever come back'. The train goes to 2046, wherever that is, or perhaps the train is 2046. In the film, 2046 is the number of the room Chow rents in his hotel in the sequences set in 1966-era Hong Kong. 2046 is also the year when China will be allowed to make changes to the way that Hong Kong is organized, fifty years after the British colony was formally returned. Protests calling for the democracy provided in the 1997 handover to China led to the Occupy movement of 2014, decades in advance of 2046. The train is the closed 'vehicle' by which all these layers of meaning are connected and it's also the place in which memory happens or is regained. This stylized set of connections depict the afterlife of the closed world in the last remaining Communist nation, a condition in which it is not so much dead as undead, dead and alive at once.

The global village

One of the products of the Cold War closed world was, oddly enough, the concept of the global village created by mass media. In a village, everyone knows everyone else's business and so the global village would be the ultimate closed world. Television – literally meaning 'vision from far away' – first created the possibility of a global visual culture on screen that digital networks have now brought into being. The conceptual space of television was at first the studio, often used much like a theatre. Outside broadcast and satellite transmission broke down the boundaries between the world on screen and the world we live in. The first live broadcast was in 1951, the same year as *Strangers on a Train*. On 18 November 1951, Edward R. Murrow used a split screen in his live show *See It Now* to show both the Brooklyn Bridge and the Bay Bridge in San Francisco. Now what could be seen on screen had clearly transcended the capacities of human vision, putting different spaces into the same frame, overcoming the limitations of time and distance.

For both sides in the Cold War, it came to seem that space was, to quote *Star Trek*, 'the final frontier' in abolishing the restraints on human action. When the Soviet Sputnik satellite first went into orbit in 1957, it was something that had never been seen before. The United States felt in danger of being eclipsed technologically and made a huge investment in space. A prototype communications satellite was launched as soon as 1960, leading to the successful launch of Telstar, a transnational project, in 1962. While satellites were primarily military, they had enormous effects on everyday life. Events could now be shown on television nationally and internationally in real time. The assassination of President John F. Kennedy, the subsequent murder of Lee Harvey Oswald by Jack Ruby, and Kennedy's funeral in November 1963 were the first 'live' media events to shape national public opinion in the United States.

During the four days from the assassination to the funeral, 166 million people watched TV coverage. All regular programmes were cancelled, no advertising was shown and all three networks (as it was in those days) carried news without a break. Television montaged scenes of Kennedy's career with those of his death, fixing the Kennedy legend in people's minds and creating instant icons, like the shot of the young John F. Kennedy Jr saluting his father's coffin. Kennedy's death and that of Oswald were the first homicides to be seen live. Watching Oswald being shot close-up while also seeing the murderer was especially shocking. In households with televisions, people watched for an average of 32 hours, meaning the set was on 8 hours a day. What was then exceptional soon became the norm.

The global audience, all watching the same events, using the same broadcast television pictures and coming to a collective viewpoint on them, seemed to mark a new direction in world history. The Canadian media theorist Marshall McLuhan called this collective viewing the 'global village'. He felt that 'the total electric field culture of our time' had recreated the conditions of what he called 'tribal societies' (1962). For the electronic extension of the senses had reduced space to such an extent that the world was now a village. In *Understanding Media* (1964), McLuhan announced that the world was in fact now imploding. He believed that what he called, perhaps for the first time, the 'visual culture' of modernity was being transformed into a new 'auditory-tactile' form by television. For McLuhan, television was a 'cool' medium that required the audience to do a good deal of work reconstructing and developing the message, unlike 'hot' media such as film. In short, to quote his famous phrase, 'The medium is the message'. It was the way media work rather than what they were doing that mattered. This emphasis on form did not mean that media were not important. To the contrary, McLuhan insisted that 'any understanding of social and cultural change is impossible without a knowledge of how media work as environments'.[3] In that sense, this chapter follows McLuhan in seeing different media forms as creating different worlds.

The period of the global village was, in retrospect, quite short. It extended from the death of Kennedy to the 9/11 attacks. In this period, global television audiences watched dramatic events like the first moon landing (1969), the wedding of Charles and Diana (1981), the fall of the Berlin Wall (1989) and the 9/11 attacks (2001). So in the course of just fifty years, watching a world-changing event became a routine consequence of technology, available to hundreds of millions of people who might have little understanding of how that technology works. People who were alive at the time can all recall seeing the TV broadcasts when President Kennedy was killed, or the 9/11 attacks occurred. Today, news breaks as much through Facebook, Reddit, Twitter and other such applications as it does through television bulletins.

Media no longer prize form so much as content. A book might be available as an e-book, an audiobook, a video, or a Braille text, as well as a printed volume. Broadcasting itself has largely shifted to 'narrow casting', organized around content rather than form. Broadcasting was a mass medium, in which the audience was given very limited choice over content but was able to receive the form very widely and usually free or at low cost. Narrow casting aims at specific audiences organized around preferences for content, such as channels devoted to specific sports, independent films, home decorating, and so on. The audience may be substantial but are more alike than different. Narrow casting usually has to be paid for and is often expensive. Truly mass audiences now tend to gather for ritualized media events such as the Super Bowl, the World Cup or the Oscars, whose content is not wholly known in advance but has very few variables.

Outside the Anglophone world, the single viewpoint of the global village had never seemed convincing. In 1950, the Japanese director Akiro Kurosawa caused a sensation with his film *Rashomon*, winner of the Academy award for Best Foreign Language film in 1952, among many other awards. It showed four different versions of the same event, a rape and a murder. What at first seemed a wanton crime later comes to seem very different as the versions of the story are told. Although the film 'solved' the mystery, its version of global media now seems prescient: whether by choice or not, we see a version of events that makes no effort to be comprehensive. All of us with access to social networking choose a set of media sources with which we are sympathetic, a process that media scholar Richard Grusin calls 'premediation' (2010). We are all living in our own version of *Rashomon*. Amazon tries to recommend other purchases to you based on what you have already bought, even if it's often another book by the same author. Facebook, which steers us to those of our 'friends' its algorithm thinks we have most in common with, even launched its Paper app in 2014, designed to create a 'newspaper' pre-customized to the user's interests. The end of a single media narrative is often lamented by the media themselves. Broadcasters like Walter Cronkite in the US or Richard Dimbleby in Britain are held up as lions of a lost era, in which we were all watching the same screen.

Total noise on screen

In 2013, marketing reports estimated that the average American now spends more time online than watching television.[4] Five hours a day were spent online while people watched television for four and a half hours a day. Some of those hours overlap, as people have the TV on and a computer online at the same time. Add to that an increasing amount of time on phones and tablets, plus the occasional visit to the cinema, and many of us spend more of our waking lives looking at screens than not. It's not just in the West. The

Chinese Spring Festival, a five-hour-long television extra-
vaganza, was watched by 750 million people in 2014. On a
daily basis, China now has 450 million online video viewers who watch an estimated 5.7 billion hours of content per
month.[5] For better or worse, we don't just look at the world
on screen, it's how we look at life.

These screens are enabled by two overlapping networks
of material cables, one transmitting electricity, the other
transferring information. The electricity network is often referred to as the 'grid', suggesting a uniform and even distribution of service. A major reconfiguration of the landscape
was necessary to provide electricity in the United States.
Rivers were dammed to create hydroelectricity, and as early
as 1920, 40 percent of US electricity was generated by water.
New bodies like the Tennessee Valley Authority were created to supervise them and a national network of transmission lines. For all this transformation of the landscape,
urban theorist Lewis Mumford called such provision the 'invisible city', a network without which cities could not exist
but that is largely invisible to their residents (1961).

In similar fashion, we can access the Internet anywhere
in the world thanks to the extraordinary fibre-optic cable
networks that have been laid since the end of the Cold War.[6]
Fibre-optic cable carries information as light rather than
as an electrical signal on a wire. It can carry far more information than wire and loses far less of it. There is a pleasing
symmetry to the idea that global visual culture is enabled by
a network of cables carrying information as light. In 1991,
the Internet was still dependent on the ARPAnet, the military network built to send messages in case of a nuclear

war. Today there are at least 250,000 kilometres (155,000 miles) of fibre-optic cable connecting the world. One such cable was FLAG, the Fiber-Optic Link Around the Globe. Laid from the United Kingdom to Japan it is 28,000 kilometres (17,400 miles) long but only a couple of centimetres around. These cables cross the oceans and circle continents but do not travel across land, as you can see on the interactive Submarine Cable Map.[7] From the Cable Landing Points, where the fibre optics come ashore, companies and cable networks build connections to consumers and businesses. These sites are key security installations because 95 percent of global Internet traffic uses these cables.

The information carried by these distributed networks is made visible by the ubiquitous screens of phones, tablets, televisions and computers. Cinema offered a given film at a specific location: the movie theatre or later the television rerun, just as the railway offered only certain destinations and times of travel. Today's screens are a blur of apps, notifications, downloads, updates, and other indicators such as time, signal and battery life. In the era of broadcast TV, a television set that had lost its picture was described as showing 'white noise', the title of a classic 1985 novel by Don DeLillo. Today we live in a condition that the Generation X novelist David Foster Wallace described in 2007, riffing on DeLillo, as Total Noise: 'the tsunami of available fact, context and perspective'.[8] Wallace realized that with so much content and context always available, we never feel adequate to the task of knowing what there is to know. That tsunami is breaking across our screens every hour of the day.

From a computer with Internet access, it's possible to

see the whole world, either literally using programs such as Google Earth that have mapped the entire planet, or metaphorically, given the limitless and constantly updated information available with a few clicks. The screens we look at now are close to us, seemingly private rather than public, and filled with information.

Just as the military satellite enabled the live broadcast and global village, the first people to get to look at computer screens like this were fighter pilots. Developed in the 1980s, helmet-mounted displays for fighter aircraft advanced from providing basic data to the current look-and-shoot technology. A pilot can actually select a target with their eyes and fire at it regardless of the position of the aircraft. As the website for the Eurofighter Typhoon aircraft, introduced in 2004, puts it:

> The Head Equipment Assembly (HEA) comprises the aircrew helmet and all the sub-system elements needed to display a real world overlaid picture on the helmet visor. Aircraft systems provide target and flight information which the HEA can combine with its in-built night vision enhanced outside world image and project them together on to the helmet display visor, exactly overlaying the aircrew's outside world view by means of a high speed helmet tracking system.[9]

With the pilot's helmet-mounted display, the synthesis of human and machine becomes a practical reality. The pilot flies in a visualized information field, created by the very machine he is supposedly piloting. This screen-directed vision is the paradigm of visual culture in the computer age, just as the dolly was to the railway world.

Figure 44 — Screengrab from Vision Systems International[10]

As this simulation from Vision Systems International, one of the leading manufacturers in this field, shows, this visual field is not intuitive. Just as we once had to learn how to see like a camera, a pilot must learn how to interact with the visualization of his display on which his life depends.

The information-packed view in the world of combat has now become familiar to the millions of gamers who play games like *World of Warcraft*.[11] The game is what is known as an MMORPG – Massively Multiplayer Online Role-Playing Game. There are actually four games in the 'world' in which you play. The first is a multiplayer co-operative role-playing game, during which players have to assemble the tools that they need to progress to the later combat stages. There are over 80 levels involved in this stage alone. Then you arrive at a competitive fantasy team battleground game. Next, a three-versus-three arena organized in a competitive ladder. And finally a 10- or 25-person co-operative dragon raid. If you are new to such games, the interface can be overwhelming.

In addition to all the information, the game creates a

Figure 45 – Screengrab from the game *World of Warcraft*

3-D effect. Just as the cinema required people to learn how to interpret the moving image – to learn that a train on the screen would not hurt them – so too does a gamer have to spend many hours learning how to play and acquiring 'experience' points, known as 'grinding' to players. Any new player is 'killed' repeatedly until they learn how to visualize the environment and control hand-eye co-ordination accordingly.

Despite all the complexity involved, these games are called massive for a reason: 12 million people subscribed to play *World of Warcraft* at its peak in 2010. Free-to-play games like *League of Legends* had 30 million active users in 2013. Now it is also one of the video games that have become a popular spectator sport worldwide. People watched an estimated 2.4 billion hours of video gaming in 2013 on websites like Twitch, not to mention 'live' video game contests.[12] The games are as meaningful to these audiences as sport is to others. Just as with sport, the viewer can play the game her- or himself but the professional is much better at it and that is part of the pleasure of watching. The information-saturated screen game does not pretend to depict the exterior world, as does the camera, so much as it reproduces the daily conditions of existence for so many workers today, from the car mechanic using a computer to determine what's wrong with a vehicle, to an accountant filling in tax-return software, or a call-centre staff member reading text off a screen to customers. Only the game is fun.

Cinema offered its audience a focused, clearly defined scene to look at once they had learned to see like a camera. The point was that the camera saw for us. Watching or playing a first-person shooter game is still in that tradition. In the data-filled screens of the fighter pilot, the multi-layered video game player, the phone user, or the stock market trader no such clarity is available. Expertise is required even to make sense of the screen. If this is total noise, it is not unintelligible. Rather, it requires people to become more open to the unexpected and to anticipate differently, as we saw with the gorilla video experiment in Chapter 2. Because screen-world culture involves active choices about what to concentrate on and what to do as a result, it can seem to offer a greater degree of freedom. While there is some truth to that idea, people often forget that anything done online leaves traces and can be found. All the apparent freedom comes at the cost of a high degree of control (Chun 2006). Whereas we once made a choice to go to the cinema or turn on the television, our devices now demand our attention with their beeps and trills. No less than 43 minutes a day are spent waiting for computers to warm up, download updates, connect to the Internet, and so on.

French philosopher Gilles Deleuze called such experience that of 'the society of control',[13] in which we are set limits within which to operate, unlike the previous disciplinary society that had firm rules. This is a society centred around controlling and defining key parameters, such as your credit score, your cholesterol level, your SAT score, your GPA, even your hits, likes and retweets – anything that can be quantified and have defined levels of success and failure. Some of this benefits the user, but it also places a greater burden on them. Once work was defined by hours spent at the workplace. Now email and other applications require almost constant monitoring. By the same token, the wealthy no longer own factories as they did in the disciplinary society but trade in financial securities. War involves insurgents, rebellions and uprisings, rather than conflict between nation states. People are not just worried about money, they are consumed by debt. All of these processes have been in important ways enabled by the screen-mediated society.

There are no limits to how far digital companies want to integrate themselves into our lives and even our field of vision. Consider the new technology of Google Glass. It's a wearable form of computing that looks like a frame for spectacles and can actually have lenses added to it. It consists of a frame and a computing device connected to the Internet. Google Glass is total immersion in the society of control. It is the civilian counterpart to the fighter pilot's helmet. Everything you see can be checked and recorded, both by the user and by Google. The prototype allowed users to access information in their visual field, take photographs and video hands-free, and to use applications, such as receiving

Figure 46 — Google Glass user

directions. Glass came online if the user tipped his or her head backwards sharply, in a rather disconcerting gesture. The device responds to verbal cues or to tactile scrolling. Information is displayed to the user's right. When Glass is on, or taking a picture, a small rectangle of light is visible to others on the device.

Google suggested possible uses, such as getting directions or being able to take a photograph, as in the example supplied by Google here, taken by Google X director Steve Lee. The point of the picture is that it would not be safe to use a camera while driving but Glass can do it for you. What is emerging is a total merger between what can be seen and what can be computed, for those that can afford to do so and subject to Google's priorities.

Glass made visible the privilege that such technology offers, retailing at $1,500. Partly in response to this criticism,

Figure 47 – Screengrab from Google Glass webpage[14]

Google suspended the Glass project in 2015. The important point here is not what happens with Glass itself but the new level of the society of control being created by the world of information on screen. Just as the wealthiest one percent already seem to live in a different world to the rest of us, it now seems as if there is a world 'they' can see, which 'we' cannot. It is common in research to speak of people as 'data points', and Glass made that definition into a wearable object. Because just as you may receive information from Google and other apps with Glass, Google will know from its technology where you are and what you do. This data will allow it to present ever more specific ads to you in the manner of the film *Minority Report* (2002). In the film, such targeted ads were glamorous billboards. What we're actually getting so far are low-tech text-based links. More than that, such technology means that the screen interface is now potentially ubiquitous. As it gets smaller and less obtrusive, the ability to be permanently but imperceptibly networked will constitute a new digital divide.

Like many other aspects of digital culture, Glass was marketed as offering unparalleled freedom to its users. Critics have been equally keen to point out how the new technologies exert significant control over us. In films like *Enemy of the State* (1998) or the various Jason Bourne movies, it is simply assumed that government security agencies can intercept all forms of data communications. This model fails to understand how sophisticated data surveillance has become, as revelations from Wikileaks in 2010, and then Edward Snowden in 2013, made clear. What happens today is not the Hollywood scenario in which a small group of highly empowered secret agents look at crisp visual images in secure locations. In fact, a vast army of data analysts combs through immense fields of metadata, scooped up whole from the Internet, like a super-trawler harvesting fish from the ocean. Just as such trawlers routinely amass quantities of fish that are not the target of the fishermen, so does this data collection gather immense amounts of information about people who are not officially targets.

The data produced by Glass and other similar technologies and software will be useful to security agencies as well. Tellingly, Eric Schmidt, Google's executive chairman, has written: 'What Lockheed Martin was to the 20th century, technology and cybersecurity companies will be to the 21st.'[15] That is to say, the airplane as a view on the world, created by the world of war, has been replaced by the technologically mediated screen view. Software computes for us the world that we see. Google, Apple, Microsoft – or which-ever digital giants succeed them – interpose themselves be-tween us and the world, carefully filtering what we may see and know by means of screens and software alike. The view seen by the fighter pilot has become the little private world of the phone screen carried around wherever we go. While we think of this as 'our' world, it is one that is carefully po-liced and filtered for us before we even get to see it. And the world it renders for us is, above all, the city, where we go next.

World Cities,
City Worlds

For most people, seeing the world still means first and foremost seeing our own city. Taken together, today's global cities make a world of their own. A century ago, only two out of every ten people worldwide were city dwellers. Now the global majority is urban. This mass migration has created the new global megacities – Sao Paolo, Delhi, Shanghai, Buenos Aires, Beijing, Bombay, Tokyo – that have outgrown the imperial and modern cities like London and New York. The new megacities are better understood as city regions, or metropolitan areas. It's hard to tell where they begin and end, harder still to determine accurately how many people live there. Shanghai has an official population of 23 million and an estimated 3 million migrants. The Chinese government has said it intends that there will be 30 million there by 2030, even though local unofficial estimates put the total population today at 40 million. There are currently 600 such global city regions, where an estimated 1.5 billion people live. They generate no less than $30 trillion (£19 trillion) annually, amounting to half of global GDP. The World Health Organization projects that by 2050, seven out of ten people will be city residents. Almost all that growth will come in developing countries.[1]

China, which became a majority-urban nation in 2011, plans to move another 250 million people to its cities. (Not to be outdone, India proposed adding 500 million new urban citizens a year later.) If that is accomplished, one billion Chinese people, likely to be one in eight of the world's population, will live in the new cities that are being created on a seemingly daily basis. It is often said that Rome was not built in a day. That is not true of the new global cities, especially in China. Chengdu had about 3 million inhabitants in 1990. By 2012, there were 14 million in the city, with a further 6 million in the surrounding areas. The city had an official growth rate of 13 percent that year and its exports were increasing at a rate of over 30 percent.[2]

The new global city extends beyond the older concept of city limits: it is a region in itself. Guateng region in South Africa extends across the cities of Johannesburg and Pretoria to townships like Soweto. You cannot understand Hong Kong without knowing its place in the Pearl River Delta and its relation to China's special economic areas in Guandong province. From day to day, the global city might experience low-intensity warfare that can escalate to full-scale insurgency or even civil war. These cities are intensely polluted, even toxic, especially, but not exclusively, for the poor. Global cities may present themselves as transparent hubs of frictionless commerce, but their residents often experience them as conflicted, dangerous and even haunted. These are the places from which we have to see the world today and where we learn how to see.

For if the classic city of the imperial period was highly distinctive – think of the way that Paris, London and Madrid have very different styles and atmospheres – the rapidly emerging global cities perhaps have more in common, based on the global computer network that moves money and information between them. There are the ubiquitous housing blocks and informal housing (unplanned buildings without legal access to services) on the periphery, surrounding the inevitable glass towers of the banks and the halogen-lit branches of global 'brand' shops downtown. Traffic is terrible, and the milky-white haze of smog is omnipresent. Because these spaces are key to understanding today's global visual culture, this chapter will concentrate on the ways in which cities now and in the past have shaped the way we see the world.

Rather than stressing the specificities of individual cities, I will look at how three city forms have shaped the world over the past two centuries. First was the imperial city (1800–1945), whose spectacular form nonetheless relied on keeping certain people and places out of sight. The imperial city was the place to see and be seen for those who constituted the public. That public was not everyone. It was mostly men, mostly white. In imperial capitals such as Paris, London and New York, the dandy and the street photographer observed and recorded without being seen. The Cold War city (1945–90) made being divided into its central, highly visible feature. Of course, such divides, epitomized by the Berlin Wall, made the two sides invisible to each other. Today's global city (post-1990) has inherited the centre-periphery layout of the imperial city and retains the divides of the Cold War at key global intersections like Jerusalem, Baghdad and Kabul. But it is literally erasing its own past and creating its own way of seeing. Seeing in the global city requires active self-censorship from its residents as part of a highly controlled environment, encapsulated by the now-notorious slogan of the New York Police Department: 'If you see something, say something.' Whatever there is to see must be reported and the citizen is now the stand-in for the police.

At the same time, whenever the police call on us, we must move on and accept that there is nothing to see here. All of this highly effective control is nonetheless haunted by a set of anxieties. How can the real be distinguished from the fake? Are cities still home or just another place? And in a world where everyone can know their GPS co-ordinates, do we still know where we are?

The imperial city

Let's begin in Paris, one of the most visited cities in the world today. According to the French tourist ministry, just under 30 million people a year made it their destination in 2012, dwarfing the resident population of 2.2 million within the twenty central *arrondissements*, or urban districts.[3] The city is carefully prepared to welcome them. The Eiffel Tower, built for the International Exhibition in 1889, is lit up at night. Once neglected nineteenth-century statues have been gilded. The coal smoke that once made the city buildings black has been scrubbed off. As the German writer Walter Benjamin beautifully put it: Paris was once 'the capital of the nineteenth century' (Benjamin 1999). It is now the largest museum in the world, the museum of the nineteenth century. As Woody Allen's hit film *Midnight in Paris* (2010) captured very well, many tourists come in search of a city that has long gone, whether it's the Surrealist era of the 1920s, or the Impressionist heyday in the 1870s.

Nineteenth-century Paris was a city world in which the urban observer claimed a certain cultural power by seeing without being seen. There were distinct limits to this power. Not many tourists today probably realize that the broad avenues they stroll down were widened by the city prefect Baron Haussmann in the 1860s in order to provide a clear line of fire against potential revolutionaries. Paris the museum bears little relation to its own history. It is so popular because it presents a nostalgic view of a city life that has long departed.

But then again, it was always this way. In 1855, the novelist Honoré de Balzac announced to his readers: 'Alas! The old Paris is disappearing shockingly fast.'[4] And the poet Baudelaire added a few years later: 'Old Paris is no more.'[5] The pioneering photographer Charles Marville became famous in the 1850s for his pictures of the old streets taken just before they were demolished to make way for the new avenues.

There are no people in those photographs. It was the buildings that provoked the nostalgia, not their impoverished inhabitants. Haussmann tore down the old Paris on purpose to protect Emperor Napoleon III from street revolution. The old revolutionary neighbourhoods were demolished and the workers sent to live outside the city centre in what became known as 'the red belt' (meaning radical) in the twentieth century. Many of today's global cities have

Figure 48 — Marville photograph, *Old Paris*

come to have the same layout: a wealthy core with good services, surrounded by people living precariously in informal housing. The centre is highly visible, actively put on display for consumption by tourists, while the periphery is invisible, kept out of sight to all but its residents.

Paris became known as the 'city of light', a reputation that began in the eighteenth century with the first installation of mirrored street lamps. In the early nineteenth century, gaslight made it possible to stroll and shop in the city at night. To facilitate this new pastime, the city created its Arcades, rows of shops covered over with a glass roof and provided with heat in winter. Shops began to display a new sign: 'Free entry.' Previously it had been expected that anyone entering would make a purchase; but here was the beginning of the modern practices of browsing and window shopping.

As Benjamin pointed out, it was as if all the centre of the city became an interior in which various modern types came into being. There was the woman of fashion, her changes in style and adornment carefully noted by the newspapers of the time. Meanwhile, men in business or government began to dress in black, abandoning the colourful male clothing of the eighteenth century for the frock coat. In response, the counterculture began. Those not wishing to be taken for businessmen were visibly not at work, such as the poet Gérard de Nerval who famously took his pet lobster for a walk on a leash.

Watching and observing all of this were the *flâneurs*, a word that is hard to translate. The word 'dandy' is close, so is 'gawker', so too 'idle stroller'. The *flâneur* was all of these. The modern city had opened a space for the *flâneur* by demolishing the narrow streets of the old city, driving out the poor, and creating a network of boulevards and arcades suitable for walking while observing. For Baudelaire, the *flâneur* was 'a prince, everywhere in possession of his incognito':[6] a man who gained a certain power by seeing without being seen – a very urban accomplishment. Becoming what Edgar Allan Poe called 'the man in the crowd' was the new way of seeing in the imperial city. The *flâneur* embodied male look-ing, the practice that would later be called the male gaze in cinema (Chapter 1).

As photography improved, the 'man in the crowd' way of looking came to be incarnated in the street photograph, taken without the awareness of those being photographed. This secrecy, and the realism of the resulting photograph are absolutely central to the success of these photographs. In recent years, we have seen repeated scandals as what seemed to be classic pieces of observation turned out to have been posed. The French photographer Robert Doisneau took a famous photograph of a passionate kiss in the streets of Paris in 1950, known as *The Kiss by the Hôtel de Ville*. The two lovers clinch, with the young man's arm around an ele-gantly dressed young women, who has seemingly been taken by surprise as her arms rest limply by her side. In the fore-ground, a man sits visibly watching them from the seats of a café. Indeed, it was standard practice for café seats to be arranged so that patrons could watch passers-by. The figures stand out against the hazy background, as if in a film noir. The scene is steeped in romance. Years later, when the photograph had become a poster classic and there was money at stake, two people sued, claiming to be the young couple. Doisneau was forced to admit that the picture was staged and the characters were young actors. He had taken shots in three different locations before deciding on the one ultimately used at the city's town hall. Why does this mat-ter? If we know it's been staged, it's not really urban ob-servation but street theatre. Now we might think that her arms stay down not because she's been taken by surprise but because she's not really kissing him. And another illusion is shattered – the illusion that street photography is like being there without being seen.

And what of the *flâneuse*, the woman dandy/gawker/stroller? There were women who wore men's clothing to gain this freedom, like the novelist George Sand. Such women formed a social type known as the Amazon, after the legendary female warriors of antiquity. The *Amazon* in Figure 49 was painted by Edouard Manet in about 1882. She's dressed for riding in the all-black uniform of the masculine bourgeoisie, including top hat and kid gloves. Her hair is in a tomboyish pageboy and there is none of the adornment we might expect. Perhaps it's just the cinched in waist that provides the anxious (male) viewer with a secure key to her gender. She presents us, as intended, with little to look at so that she can claim the right to look herself. There were numerous women artists who painted and drew their lives in the modern city, such as the Impressionist painters Berthe Morisot and Mary Cassatt. At the same time, one of the subjects that most fascinated the male gawker was the Parisian woman, something that has not gone out of style to this day. Recent art history has claimed that the paintings of women on their own in modern Paris would have implied to their contemporaries that these were sex workers. The term 'public woman', we are reminded, was a euphemism for prostitute. Yet there is still ambiguity.

I think of Edgar Degas' painting *L'Absinthe*, showing a woman drinking an absinthe by herself in a café (1876). There was certainly a scandal about the work at the time, allegations of degeneracy and alcoholism being thrown

Figure 49 — Manet, *Amazone*   **Figure 50** — Degas, *L'Absinthe*

around by the usual suspects. The woman depicted was Ellen André, a well-known popular actress, who also appears in paintings by Renoir. Absinthe was a powerful drink, alleged to induce hallucinations and much favoured by the bohemian set. André has a full glass in front of her. We cannot know if she will drink it or not, whether it is even her first ever, or one of many that day. She's very fashionably dressed in white, with an elaborate hat. These are not street-walking clothes. She is notably by herself. There's no interaction with the man seated next to her. Her thoughts are not available and her expression is blank. You could say we are free to look and imagine what we will. You can also say that she is not knowable and has a precise degree of independence. She is not at home, the Victorian angel-by-the-hearth, nor is she demonstrably a sex worker.

What makes the period still seem contemporary is

precisely its fascination with such leisure and consumption, the primary activities at the centre of so many global cities today. The imperial city made spaces for these activities and pushed those who had nothing to do but work to its margins. Impressionist paintings showed men and women picnicking, boating, flirting, at the opera, at the café, at a concert, at the ballet, and so on. Work was elsewhere – except for the sex workers, performers and restaurant staff that made this all possible. The paintings seem now to embody the very idea of a city of light, with their bright colours and flickering brushstrokes.

One well-known piece of art history trivia is that the Impressionists were unpopular in their own time. It is less well remembered why. The name Impressionist was not a compliment. In traditional oil painting, the artist made an 'impression' of the scene they were going to render, a quick sketch that would serve as a guide in composing the finished work. So for viewers of the time, the Impressionists were presenting unfinished sketches as proper painting. Just as some people today look at abstract painting or conceptual art and think that it's not really art, so too did nineteenth-century critics see the now-loved paintings as being half-done at best. The Impressionist style has come to seem to depict the urban observation of the *flâneur*, catching a glimpse of what was going on out of the corner of his eye as the crowds and traffic hurry by.

The artists were well aware of what they were doing and claimed to be depicting colour in accord with the discoveries of nineteenth-century science. The implication was the traditional ways of painting no longer depicted the modern city effectively. The bright modern colours stem from another technical change. Traditionally, an artist would cover the entire canvas with a coloured ground, whether red, grey or brown. It was designed precisely to mute the force of the colour in the actual painted scene. The Impressionists painted on white ground and so their work 'pops' off the wall, recognizable from a distance. While an art critic of the day would have seen the colour as being out of control, the work of degenerate bodies, we now see these paintings as the highpoint of modern beauty. What appeared hectic and world-changing in the nineteenth century is soothing and calming today.

By the same token, the Arcades were the precursor to the ubiquitous malls that you can now see in global cities, from Johannesburg to Shanghai. The mall is a covered area for the purposes of consumption; it is more likely to be artificially than naturally lit. Some malls, like the one inside Caesars Palace in Las Vegas, go to great lengths to create a 'natural' lighting effect that tricks the brain into thinking we are outside. Crowds flock to the 'outside' seating within the mall to watch the 'sunset' every two hours, while the 'inside' seats are all deserted. Outdoor spaces like Times Square in New York and Causeway Bay in Hong Kong are now illuminated with these 'natural' lights, giving the uncanny feeling of being in daylight at night. The city of light is now the global retail blueprint. A certain degree of luminosity seems to encourage and enable us to spend on things that we don't really need.

If the once legendary cafés of Paris, each with its own character, have become the globally uniform pastiche

Starbucks, the nineteenth-century dandy would nonetheless recognize them and the malls of which they are a part. The major addition has been the merger of the cinema with the mall, where multi-screen cinemas are one of the anchor businesses (Friedberg 1994). The work of watching the world, once the specialist task of the dandy/gawker/stroller, is now made available for us for £10 in luxury seating with a holder for a vast sweetened drink. Unseen observation is just another commodity in the global city.

What was less visible in the city of light was that Paris was not just the capital of the nineteenth century. It was the capital of the French empire, from Africa to East Asia and the Caribbean. From the capital 'N' that marks many buildings and bridges (N for Napoleon, the first Emperor, also used by his nephew Napoleon III), to the Louvre, filled, then as now, with booty from imperial wars, for example Egyptian sarcophagi and Greek sculptures, and the ubiquitous sugared coffee grown in the colonies – Paris could not be understood except in the context of empire. And the Parisians knew it. In 1832, after the failed revolution of that year, later celebrated by Victor Hugo in *Les Misérables*, a French journalist observed:

> Every factory owner lives in his factory like a colonial planter in the middle of his slaves, one against a hundred; the uprising is to be compared with the insurrection at Saint-Domingue [Haiti].[7]

The nervous writer visualized what he called 'the possessing class' as colonizing those without resources, and feared that a successful revolution, like that of the former slave colony

Saint-Domingue, which became Haiti, could only be a matter of time. Indeed, in March 1871, the Paris Commune took over the entire city and created what they called a 'free, autonomous and sovereign' space. French troops swept back in a few weeks later, killed an estimated 25,000 people and restored central government. That regime lasted as the Third Republic, right up until Hitler's invasion in 1940. The pacified imperial city became the backdrop to the Impressionist paintings and other nostalgic trappings of today's museum Paris.

Outside the inner core of today's Paris is another city, four times as large. Once the home of the white working class, it is now where the descendants of France's empire mostly live. Known as the *banlieux* – the suburbs – here are the homes of the immigrant populations from North and West Africa, the Middle East and Asia. Unemployment is high. Consequently, so is crime and drug use. Policing, which is typically discreet in central Paris, is visible everywhere. Transport is difficult, since the metro ends at the 'gates' of Paris. From there you must take a bus or an additional light rail to the *cités*, or 'cities', as the huge public-housing blocks are known. There is none of the charm of inner Paris here, no little squares and cafés, just block after block of towers and very little to see or do. Paris's layout was designed to keep this separation as intact as possible. A separation by ethnicity has replaced the old separation by class.

Divided cities

During the Cold War, certain cities became separated and divided in ways that could not be ignored. So if Paris was the paradigm of nineteenth-century imperial cities, Berlin was the classic city of the military-industrial complex (1947–90). Governed by the victorious war powers, it was divided by the monumental Berlin Wall from 1961 to 1989. The city was visually split in dramatic fashion, equivalent to the clarity of the U2 photograph of Cuba (Chapter 3). In Berlin, no visual work was required. You simply lost the ability to see the partitioned space. Although the Cold War is long over, divided cities are recurring and reviving in critical areas of global counterinsurgency, from Baghdad to Jerusalem and Kabul.

At the end of the Second World War, in 1945, Berlin was divided into four sectors, one controlled by each of the Allied powers, namely Britain, France, the United States and the Soviet Union. On 13 August 1961, astonished Berliners awoke to find the former GDR (East Germany) building a wall between its sector of Berlin and the Western sector. For nearly thirty years, the wall served as both a symbol and the reality of Cold War separation. It was 140 kilometres long, 3.5 metres high and surrounded with mines, dogs, lights and other security devices. It cut through neighbourhoods, separated friends and families and provided the ultimate visible symbol of the Cold War. Although East Germans were able to learn about the West from television and radio broadcasts, their personal movement was directly limited by the Wall. Subway lines eerily missed stations on 'the other side'.

Before the construction of the Wall, as many as 3.5 million East Germans are estimated to have defected to the West. The Wall made that almost impossible, although some 600 people died in the attempt to cross it.

The official position of the ruling Socialist Unity Party (SED) in the GDR was that 'there exists no objective political or social basis for opposition to the prevailing societal and political order'. In other words, no sane person could oppose what the SED named its 'comprehensive form of democracy'. Any opposition was therefore wrong and carefully monitored by the Stasi (Ministry for State Security). You can now visit their vast headquarters in East Berlin, which has been preserved as a museum. The Stasi had no expectation that their monitoring of the people would lead to better behaviour, they simply wanted to control them. So they regulated and determined the boundaries of acceptable behaviour and held their fellow citizens accountable for any breaches of those boundaries. In the displays of Stasi equipment at the museum, you can see a 10-megabyte hard drive that was used to store information. As it is of late 1980s origin, the disk is twelve inches across and six high. It is surrounded with a display of the five-inch floppy disks that were also used, as well as the mounds of paper generated by the constant surveillance. The display hints at a future that was yet to come – in 1989 – but is now all around us. Indeed, the GDR made immense efforts through its Robotron company to keep pace with the digital revolution in the 1980s, which contributed to its financial collapse.

The material fact of the Wall created a social fact of segregation. This new fact had to be learned and was

Figure 51 — Checkpoint Charlie

constantly emphasized by signs. Above is the famous sign at Checkpoint Charlie in Berlin, seen in so many Cold War-era movies like *The Spy Who Came in from the Cold* (1965). It marked the boundary of the American sector of the city with East Berlin. It was one of only two crossing points for foreigners to enter the GDR in Berlin and the only one that members of the armed forces could use. It became mythologized as the location of spy exchanges and other intrigue. There were no such signs on the GDR side because citizens were forbidden to approach the Wall.

In 1963, President John F. Kennedy spoke at the Wall and famously said: 'Today, in the world of freedom, the proudest boast is "*Ich bin ein Berliner!*". . . All free men, wherever they may live, are citizens of Berlin, and therefore, as a free man, I take pride in the words "*Ich bin ein Berliner!*"' (I am a Berliner). His meaning was clear: that Berlin was a symbol of the freedom claimed by the United States in the Cold War and it was willing to defend the city as if it were its own sovereign territory.

There was an obvious contradiction that Kennedy's Soviet opponents did not fail to point out. Almost all American cities south of the Mason–Dixon line, which separated former slave-owning states from free states, were still divided in 1963. Signs in the street indicated who could go where and who could do what. Only these were citizens of the same country, divided by the colour line (Abel 2010). Across the South, you could see signs indicating that one rest-room, water fountain or entrance was for 'whites' and another for 'coloured'. Such signs, and the law they indicated, divided these towns and cities as precisely as any wall. Crossing the line was often dangerous. The Civil Rights Movement challenged segregation by undertaking highly visible actions in which the unity of the nation was posed against the divided reality of segregation. In the small town of Greensboro, North Carolina, students trained in non-violent civil disobedience sat in at a lunch counter in Woolworth's department store on 2 February 1960.

The students concerned were Joseph McNeil, Franklin McCain, Billy Smith and Clarence Henderson. On the first day, they sat for an hour at the end of the day without being served. On the next day, when this now famous picture was taken for the local newspaper, they sat for an hour and a half. The students were well dressed and conservatively groomed so that no objection could be raised about their personal appearance. During the sit-in, they sat quietly and

Figure 52 — Moebis, *Woolworth's sit-in*, Greensboro, NC

often studied (Berger 2010). In the photograph, you can see an African-American waiter or bus boy studiously ignoring his peers, as did all Woolworth's staff. The hope was that the sit-ins would simply go away. Instead, they spread across the South.

The sit-in was a targeted tactic. The activists were asking only to be allowed to spend their money. By making it clear that the segregated South preferred prejudice to business, the action made the colour line indefensible at this particular place. By making the refusal to accept money apparent, the sit-ins created a new link between what was sayable and what was visible. Segregated businesses were not discussed, they were simply a 'fact', part of the common sense of race. The apparently simple act of sitting at a counter to ask for service was unthinkable. Once this was challenged, the lunch counters were integrated within weeks, though many arrests were made as well.

The relatively simple gesture of the sit-in again raised the stakes of the issue to another level. The provocative slogan 'States' Rights' that claimed segregation to be a matter for local decision making (by the white minority) was now countered with Martin Luther King's repeated reminder that the Declaration of Independence held it to be self-evident that all people are created equal. For John Lewis, then an activist with the Student Nonviolent Coordinating Committee (founded in 1960), and now a member of the House of Representatives, the experience of the sit-ins was one in which 'democracy was lived as a reality'.[8] Others spoke of having their soul cleansed. So successful was the countervisualizing of segregation that it has now been absorbed into the national narrative of the United States. In the 1960s, civil rights activists were described by the FBI as Communists, terrorists and worse. Now, the Civil Rights Movement is seen as a symbol of the American capacity to overcome hardship and create a more perfect union, as set out in the Constitution.

If we compare the history of South Africa, where an even more rigid system of segregation, known as apartheid, or separate living, was enforced, the contrast is striking. It was not enough to simply make social and commercial segregation visible. Apartheid reached into every corner of South African life with extraordinary determination. Racial distinction was hypervisible and violently enforced.

The sign below, now preserved in the District Six Museum in Cape Town, indicates the level of separation that

Figure 53 – Sign in District Six Museum, Cape Town

apartheid tried to maintain: here a rest-room is designated not just for 'whites', as in the United States, but specifically for 'white artisans'. Based on the ownership of land, and the resulting agriculture and mining, apartheid was central to the country as a whole in a way that Southern segregation in the United States no longer was by the 1960s. Apartheid was in fact made more rigid by the creation of the whites-only National Party in 1958 and the creation of a Republic in 1960, ending all formal ties with Britain.

On 21 March 1960, the police opened fire on marchers in Sharpeville, protesting the Pass Law by which all Africans (as the indigenous populations were known under apartheid) had to carry a passbook detailing their identity, residence, tax status and more. When the shooting was over, 69 people were dead and 180 injured, reproduced in newspapers around the world. In most places, a massacre of this kind would have produced noticeable change. The terrible Birmingham, Alabama, church bombing in 1963 killed four

young girls and is widely thought to have been a contribu-
tor to the passing of the Civil Rights Act in 1964. Sharpeville
changed nothing at the time. It convinced many black South
Africans of the need for armed resistance as the only way
forward. South Africa at that time was based on legalized
white supremacy and until that came to an end, no single
event was likely to change it. The visible distinction of 'race'
overwrote all other issues and priorities in defence of the social hierarchy.

The apartheid regime nonetheless did everything it could
to spare its white residents from being confronted with
alternatives. After Nelson Mandela was arrested in 1962, only
two photographs of him were released during the twenty-
seven years of his imprisonment. The non-white population lived in separate locations that were hard to reach. Yet
there were all kinds of professional and personal relation-
ships between the supposedly separate ethnic groups. The
white minority supervised an African labour force and had
Africans doing domestic work and child care. If we compare the work of (white South African) photographer David
Goldblatt (b. 1930) and his (black South African) counterpart Ernest Cole (1940–90), these contradictions were in-
terestingly visualized. Both photographers seem to adhere
clearly to the modernist aesthetic of 'show not tell'. Their
work is observational, not prescriptive. Both were nonetheless considered shocking, to different degrees.

Goldblatt's book *Some Afrikaners Photographed* (1966)
caused considerable controversy in the country at the time.
Afrikaans-language media (the language spoken by Dutch-
descended Afrikaners in South Africa) were incensed: 'blood

Figure 54 – Goldblatt, *A Farmer's Son With His Nursemaid*

will boil' was one typical headline. The scandal is, however, all in the implications.

In this photograph, taken shortly after Nelson Mandela had been sentenced, we see a moment of everyday life that both epitomized apartheid and shows why its proponents wanted to keep it invisible. The picture is about land and power. The two figures are divided by race, gender and access to power. The child is identified as the farmer's son, presumably the heir to all the land we can see. He stands confidently, directly addressing the camera in a classic *contrapposto* stance. The woman is identified as his nursemaid and it is possible that she did literally nurse him. She turns her body away from the camera and her expression is hard to read – a mixture of deference, recognition and curiosity. The boy seems to dominate the adult. He places a hand on her shoulder, while she reaches behind her to touch only the back of his ankle. The furtive touch offered by the carer to the farmer's son takes place against the symbolic backdrop of a barbed-wire fence. The fence marks the limit for people without authority and animals alike. The nursemaid is seated so that her head is below the fence, while the young farmer stands above the (colour) line.

If Goldblatt explored everyday apartheid at 'home' on its farms, Cole went into the divided cities, where it was contested and enforced every day. He learned his skills on the now-legendary intercultural Johannesburg magazine *Drum*, where he started as a teenager in 1958. Cole was able to get classified as 'coloured' rather than 'black' despite his dark skin because he spoke Afrikaans. This classification allowed him sufficient freedom of movement to pursue his photographic project, documenting the newly onerous apartheid regime, taking photographs of street life until 1966 when he smuggled them out of the country. His book *House of Bondage* was published in the United States in 1967. It was immediately banned in South Africa, where his photographs were not exhibited until 2010, some twenty years after his death. If Cole was indeed the 'man in the crowd', he worked in a context where all people were not equal.

This photograph from the book shows how the enforcement of the Pass Laws in the streets of Johannesburg centred around the exchange of looking. An African policeman detains a young African man, as many people watch. By today's standards, the policeman is using minimal force because he expects to be obeyed. Passers-by, some African, one white woman, two apparently 'coloured', all regard the encounter with different degrees of engagement. The

Figure 55 — 'Pass Laws', from Ernest Cole's *House of Bondage*

white woman seems not to be troubled, while the African woman closest to the scene looks directly at the arrest with apparent concern. To the right stands a white man with a moustache, apparently supervising the arrest. It is not clear whether he has actual authority over the policeman, as his superior officer, or just symbolic authority as a white man in the apartheid regime, but it is clear that he feels in charge. He stands directly in front of a newspaper sign from the Johannesburg *Star*, which reads: 'Police Swoop Again.' Even the poster on the pole at the right seems to be looking at the encounter. Indeed, taking this picture led to Cole's own arrest. When apartheid finally collapsed, there was shock and dismay among many whites when the Truth and Reconciliation Commission made visible what had been unseen by them for so long but had always been there – the violence of racial classification and separation. Like the woman in Cole's photographs, white South Africans had simply passed it by.

Today there are black Presidents in both the United States and South Africa. Clearly, much has changed. South Africa's artists are as much part of the global art world as Johannesburg is part of global capitalism. The Johannesburg Stock Exchange is worth more than all the rest of Africa's stock exchanges put together. All Africa's top 100 companies are in South Africa. Certainly that speaks to the economic weaknesses of Africa, but it also shows that the transfer to majority rule did not, as was so often predicted, lead to economic collapse in South Africa. Much has been done to extend electricity, water and mains sewerage to the majority population. But immense wealth gaps between rich and poor and black and white persist. Despite some notable exceptions, the rich are still mostly white and the poor mostly black. The net worth of the average white household is just under a million rand (£58,000). Their black African counterparts are worth some 73,000 rand (£4,200).[9]

One of South Africa's best-known new artists is Zwelethu Mthethwa, who works in the townships, mines and farms, depicting those places where black Africans still largely live and labour. Mthethwa came to international attention with photographs like the one in Figure 56 (2000), showing the interior of a township house. Many visitors see the outside of townships, at least on their way to and from the airport. Few are invited inside. Mthethwa's work shows how the township residents take pride in their homes and do their best to decorate them, using coloured pages from magazines and newspapers. James Agee's photographs of impoverished white

Figure 56 — Mthethwa, *Interior*

households in the American South during the Depression show a similar wallpapering with newspaper. The layer of paper also serves as insulation. In the Mthethwa image, the necessity and labour of fetching water predominates. The small space is filled with buckets and other water-holding devices. The shelf on the wall is not level. The floor is made of brick, so the dwelling is not intended to be temporary. The seated woman is no longer formally contexualized by racial difference. She is alone in her space. Clean and well presented, visually she refuses to be a victim.

It is also the case that the expectations she might have had when freedom came in 1994 have not been met. Ending the formal classifications of apartheid has not erased the colour line. There are now some white South Africans living in informal housing like this, but very few. As a result,

South Africa has developed a major problem with crime, leading to a highly policed and segregated urban space. Personal wealth is now indexed by the number of keys a person carries (Vladislavic 2009). The few strands of barbed wire in the Goldblatt photograph have been replaced with high walls and rolls of razor wire. Surveillance cameras, dogs and armed guards are everywhere in white neighbourhoods, while townships are brightly illuminated by very tall street lights, set high to prevent theft of the fixtures. In an extraordinary development, Mthethwa himself has been accused of murdering Nokuphila Kumalo, a black South African woman, said to be a sex worker, in Woodstock, a township near Cape Town. The case is set to come to trial in 2015. Kumalo was exactly the kind of person that Mthethwa photographed. Whoever killed her, her death shows the limits of 'freedom' in the global city for the global majority.

By the same token, although the Cold War is over, wall building is back in fashion worldwide, from gated communities to national borders. States have reverted to walls of exclusion. Most notably, Israel is now divided from its Occupied Territories in the West Bank by a separation wall that is 8 metres (26 feet) high. The wall was first announced in stark terms in 1994 by then Prime Minister Yitzhak Rabin: 'We want to reach a separation between us and them.' Construction began eight years later in 2002 and the wall is currently over 700 kilometres (430 miles) long. It roughly follows the 'green line' that divided Israel from what are now the Occupied Territories in 1948. However, it extends from 200 metres to 20 kilometres into that space in order to 'protect' Israeli settlements and other interests. The wall is

Figure 57 — *The Separation Wall*, Israel–Palestine

redrawing the international map on the ground and its path is often confusing.

It has gradually been covered with graffiti and posters and is becoming an uncanny reminder of the Berlin Wall. In the photograph above, someone has written Kennedy's famous quote '*Ich bin ein Berliner*' on the separation wall. Today, 'Berlin' is in the West Bank, the tagger suggests.

However, the Berlin Wall marked a clear dividing line, known to all. Today, in addition to the physical barrier created by the wall, separation takes place on many levels, as part of what the Israeli architect Eyal Weizman has dubbed the 'politics of verticality' (2007). In this politics, separation is more crucial than ever, but it extends from underground to the sky, dividing domains such as water supply, air traffic control and mineral rights. Even the access to roads is divided according to whether you live in Israel or Palestine.

In the graphic in Figure 58, created by the Beirut-based collective Visualizing Palestine, we can see that many roads are accessible only to those with orange Israeli licence plates, even in the Occupied Territories. Such licence holders also have access to bridges and tunnels that connect Israeli settlements to the east of the separation wall. By contrast, road blocks, checkpoints and even trenches prevent or restrict travel between the different Palestinian enclaves for those with Palestinian green and white plates. The checkpoints are mobile and can appear at any place and at any time. Israelis and Palestinians have become invisible to each other. Divided cities are always a tiny minority of all cities, but they express the key tensions of their time by means of the highly visible physical barriers that make some places invisible to others.

The global city

The global city is a space of simultaneous erasure, division and expansion that is hard to see and harder to apprehend. Old divides are erased, only for new ones to be built. Familiar spaces disappear, to be replaced with endless new space that is hard to differentiate. Seeing becomes a complicated matter, closer to the visualizing of a battlefield. We have to remember what was there before, try and take in what has been put in its place and keep up with the pace of change. While there are fewer formal barriers, these cities are clearly not equal for all.

Figure 58 — Visualizing Palestine, *Segregated Road System*

Recent thinking on memory and place has made much use of French historian Pierre Nora's concept of 'places of memory' (*lieux de mémoire*),[10] suggesting that particular physical locations are a key element in the mental construction of place in general and memory in particular (2006). While this may apply well to relatively stable and long-lived nations, such as France – where it originated – it does not seem as useful for rapidly changing cities like Berlin, let alone the global cities. Memory comes to seem like yet another first-world privilege, odd as it may seem. For while the disasters of Europe are well remembered, those of Africa are far less well attested to.

That is not to say that memory remains the same. Today, the Berlin Wall has been almost entirely dismantled, its former course marked only by a line of cobblestones. Like many other global cities, Berlin is in the process of radical transformation. If you have not been back to the city for some time, you might exit a familiar U-Bahn station only to feel a radical sense of disorientation as an unknown new building or structure appears – am I in the right place, did I forget what it looks like, or has it changed beyond recognition? Memory is being changed.

In 2006, I saw the Palace of the Republic in the former East Berlin being dismantled as part of the active erasure of the Communist era. Built in 1976 to celebrate the Socialist state, the Palace was the location of Party congresses and other such occasions. Now, a reconstruction of the eighteenth-century castle of the former Hohenzollern monarchs that preceded the Palace in the same space is underway. It will be an odd building. It copies the old palace exactly on three sides but there will be a contemporary glass wall on the fourth. In theatre, the 'fourth wall' is the name given to the illusion created when we look at actors on stage, as if they are simply carrying on their lives behind a transparent wall. Now the global city incorporates zones of fake space. While these zones provoke controversy at first, they inevitably become accepted and blend into the cityscape. For the fake is emblematic of globalization, and today, it is often difficult to distinguish from the so-called 'real'. For example, 'fake' Chinese watches use the same Swiss watch movement as the brands they are imitating (Abbas 2012). Such a fake watch is materially the same as the real one but lacks the cultural prestige of the 'real' brand. The fake castle links the history of kings and queens to the architecture of the global shopping mall. It is almost history, but not quite.

At Tianducheng, a gated community near Hangzhou, China, there is a 107-metre (350-foot) high replica of the Eiffel Tower among 30 square kilometres (12 square miles) of Parisian-style architecture. If British is more your style, Thames Town (Figure 59), close to Shanghai, offers cobbled streets and Tudor houses around market squares. If this is clearly a fake in one sense, you can nonetheless live in these houses in comfort. As the photograph suggests, the wealthy clients who buy these homes are rarely actually in residence. The same has happened to upscale districts of all global cities. In London, for example, wealthy neighbourhoods like Belgravia and Knightsbridge are increasingly empty on a daily basis because the owners of the prime real estate are elsewhere, in another global city. Those the economist Joseph Stiglitz called the one percent (2011) now live globally in a fake world of nineteenth-century European and

Figure 59 — Thames Town, China

mid-twentieth-century American urban living that no longer actually exists. At the same time, many of those aspiring to be among of the one percent (one poll showed 42 percent of Americans believe that they are or will be in the one percent) are carrying fake Louis Vuitton bags and wearing fake Rolex watches. This peculiar and unequal mirroring of fakery epitomizes the way of seeing in the global city.

When trying to see the new 'fake' but all-too-real global city, it helps to use methodologies from science fiction. In his novel *The City and the City*, China Miéville described two cities that exist in the same footprint (2009). One street might be completely in Ul Qoma, the next in Beszel, another might be partly in both cities. In order for a citizen to negotiate this space, in Miéville's world, it was necessary to learn how to 'unsee' spaces from the other city. Non-residents, especially children, find this hard to impossible, and the unseeing was carefully policed by a mysterious force called Breach. In the novel, to see what should be unseen results in a 'breach' that would be punished by disappearance. The global city is there and not there, requiring that we notice and ignore it at once.

In China, a massive new urbanization is transforming the built environment in ways that make the rebuilding of Berlin seem low key. It has two registers: showpiece districts, built to impress international visitors and local officials, contrast with the endless places of work and residence for the local population only. In China, they call the new residential towers 'handshake blocks' because they are placed so close together, it's as if people could reach across from one tower to the next and shake hands. More formally, the German photographer Michael Wolf calls the phenomenon 'the architecture of density'.

Figure 60 — Michael Wolf photograph from *The Architecture of Density*, Hong Kong

By taking photographs of these new buildings in Hong Kong without including any way of judging where they begin or end in the frame, Wolf has found a new modernist aesthetic at work in what appear to be simply utilitarian spaces. Such blocks create an implied visual clash with the reflecting glass towers of global capitalism. Usually the residential spaces are safely out of sight of the commercial areas. The glass towers are built to illustrate the presumed transparency of global capitalism. As we discovered during the financial crisis of 2007, and afterwards, they conceal more than they reveal. In fact, the glass only allows those within to see out. These one-way-mirror buildings are the built environment of a world order that 'unsees' its supposed citizens. Meanwhile, the residents of the handshake blocks can barely see anything out of the small windows of their apartments, looking out on a forest of other such blocks.

The artist Sze Tsung Leong has set about documenting the rebuilding of China's popular neighbourhoods. Picturesque low-rise buildings repeatedly give way to massive modern developments, uniform in style and appearance, set against the permanent air pollution of industrial China. His photographs at first were intended as a parallel to European precursors like Marville, who had documented the transformations of nineteenth-century Paris (see Figure 48).

Like Marville, Sze's photographs rarely show people, concentrating instead on the buildings. Before long, Sze felt that his work was very different from European nostalgia. His photographs show instead 'the absence of histories in the form of construction sites, built upon an erasure of the past so complete that one would never know a past had existed.

Figure 61 — Sze Tsung Leong, from *History Images*

And they are of the anticipation of future histories yet to un-
fold, in the form of newly built cities.'[11] There is certainly
unseeing of the past but it is not yet complete. In Tash Aw's
2013 novel *Five Star Billionaire*, set in Shanghai, the characters are all trying to come to terms with the intense pace
of change in China. 'Every village, every city, everything is
changing,' a young woman says. 'It's as if we are possessed
by a spirit – like in a strange horror film.'[12] As we shall see
later, this is precisely how change in the global city is visualized in horror films, when the past itself becomes the haunting spirit that 'breaches' the seamless present.

In Shanghai, there is a visible clash of empires – the old
colonial empire and financial globalization. Taken together
they visualize the cross-hatching of what political scientist Martin Jacques has called the 'contradictory modernity'

revealed by the rise of China (2011). Until recently, there was a consensus among all those involved that there was only one way for a nation to become modern – the Western way. To be modern meant having a representative democracy, free markets and a civil society with freedom of expression, and so on. China's ascent has shown there are at least two ways to be modern. China has combined a very strong state which tightly limits personal freedoms with managed economic liberalization. According to Jacques, what is central for China is its distinct culture and long history of civilization rather than a set of 'self-evident' principles. And so now we are working out this contradictory modernity. Either one side is right and the other wrong, or there are multiple ways to be modern.

On one side of the Yangzi River in Shanghai, there is Pudong, the new commercial and financial district of the city. A forest of spectacularly designed skyscrapers confronts the viewer as a wall. Pudong classifies Shanghai as a

Figure 62 — Pudong, Shanghai

global city and separates those who work there, and above all those who own land or buildings there, from other, lesser beings. It is the city in the city.

It demands respect for its sheer scale, newness and spectacle. Who knows what actually goes on in these endlessly photographed buildings? On the other side, the old colonial waterfront, known as the Bund, survives in external form. Shanghai was opened to the West following the Opium War of 1839–42, in which Britain fought the Chinese empire for its right to trade opium to the vast Chinese market. Fortunes were made. The Edwardian headquarters of the former opium dealers Jardine Matheson (long since become respectable) still stares out at Pudong, although no sign indicates its past history.

The building has become a fashion boutique, oddly entitled the House of Roosevelt, selling fake luxury goods. Every night, neon displays on the Pudong buildings put on a show for the watching crowds lined up along the colonial

Figure 63 — Jardine Matheson mansion, Shanghai

Bund. The message, here and elsewhere in China, is very clear: our way is winning.

The built environment of the city feels less real than its electronic network. Ghosts and spirits are perceived in electronic and digital media as a means for people to explore their anxieties about the seemingly endless transformations of their everyday lives. In film, old powers retain their force. New media can be controlled and manipulated from within. The classic of the genre remains *The Matrix* (1999) and its Platonic cautionary tale about computers and a digitally simulated city. The Wachowski brothers, who wrote and directed the film, wanted us all to remember Plato's ancient rubric about the deceiving nature of appearances. In the film, computer code creates a fake world that manipulates us into believing that we are free, while our bodies are in fact serving as batteries for the Matrix.

Learning how to see the simulated machine world of the Matrix is the key to resisting it. In perhaps the best sequence

Figure 64 — Still from *The Matrix*

in the film, Morpheus (Lawrence Fishburne) offers Neo (Keanu Reeves) a choice. He can take the red pill and see the world for what it is or take the blue one, forget what he has heard and go back to everyday life. But Morpheus insists, 'You have to see it for yourself'.

In the Hong Kong horror movie *Tales from the Dark* (2013), modern media like a mobile phone or a CD all turn out to be haunted. The scariest character in the film is the ruthless city of Hong Kong, haunted in advance as it is by the knowledge that it will return to full control by China in 2046. At present, Hong Kong is in China but not of it. When you enter the region, your passport is not stamped. It's as if the city is in no man's land. In *Tales from the Dark*, the voice-over expresses this view clearly: 'Humans. Ghosts. Everyone is searching for the way home'.

In the world made by global cities, it is becoming harder to find a home. In the wake of the Spanish economic crisis, the popular slogan 'I'll never have a fucking house' can be seen on walls across the country. Many Californians have complained that due to escalating rent and home prices, San Francisco has expelled everyone who is not wealthy to remote hinterlands and created a theme park of technology. London is not only driving its less wealthy citizens out by dint of astronomical rents and house prices, it is observing them. Britain now has over 4 million closed circuit TV cameras, nearly all owned by private companies. Local councils in London claim to have just 7,000 cameras but note with pride that this is far more than the 326 in Paris. Today 95 percent of all Metropolitan Police murder cases in London make use of CCTV footage.[13] While some cities are reshaping themselves under close surveillance, others like Detroit are collapsing. Detroit is lacking over 40,000 street lights. Whole districts of the city are going dark. Within the 360-square-kilometre (140-square-mile) city boundaries of Detroit – a city expanded by the automobile which made it famous – the area of now-vacant land is the size of San Francisco. These patterns are linked. In Detroit, San Francisco is 'unseen' (or perhaps unseeable), and vice versa.

In Paris, the two spaces are concentrically wrapped around each other, yet the wealthy, mostly white centre 'unsees' its poor, mostly black and brown suburbs. The pasts of the global cities have been erased, invisible and yet still remembered, at least for now. When French writer Michel de Certeau wanted to imagine how to see everyday life in the 1970s, he went to the top of New York's World Trade Center and looked down at the city around him (1984). You cannot go there any more, literally or metaphorically.

Map world

Seeing the cross-hatched, divided, disappearing, expanding global city is not so simple any more. But to see where we are, we turn back to our screens. An odd legacy of the Cold War has been a new way to map. After the Sputnik satellite created a sense of panic in Cold War America, part of the response was to create a set of satellites that would allow for precise positioning on any part of the earth's surface, known as the Global Positioning System, or GPS. Launched over a twenty-year period for the correct targeting of nuclear weapons, the GPS system was fully realized only after the end of

the Cold War in 1994. Owned by the United States government, it comprises 24 satellites, whose use was gradually extended from military to civilian. A GPS receiver calculates
its position by timing the reception of the user's signal from
four of the orbiting satellites. Millions of people now carry
such devices in phones and other personal data organizers.
For the first time in history, those who do have access to
GPS can precisely locate themselves without requiring technical skills. Devices designed solely to access GPS are not
reliant on phone service and so it's possible never to be lost.
Or at least to know where you are, even if you're not sure
where that is.

To close that gap, a variety of mapping services have appeared, from navigation systems for vehicles, to free services like Google Earth and Google Maps. Google Earth is
a massive database that is rendered as if it were a seamless
visual representation. Google Maps (and other such applications) is designed to be of practical use, offering directions, detailed indications of what each building at a given
location does, and even the ability to 'see' a specific street
via the Street View service. And if even this is too complicated, the software will give verbal directions. Google sends
vehicles equipped with automatic roof-mounted cameras to
photograph every street they can access. Using this function
allows a viewer to see what their destination will look like
before arriving, which can be useful in unfamiliar locations.
You can also simply browse for pleasure to see what certain
places look like. Some worry that thieves use Street View to
target desirable properties.

Google Earth and Street View use a process called

Figure 65 – Valla, *Postcard from Google Earth*

'stitching' to link enormous numbers of individual images into what appears to be a continuous depiction. At certain points, the illusion in these softwares breaks down because of a glitch in the system. The artist Clement Valla has made locating such errors into an art form, which he calls *Postcards from Google Earth*. These failures to render create images that are nonetheless oddly familiar because they look like the CGI-created disasters that litter today's multiplex cinemas.

As Valla puts it on his website, Google Earth is

> a new model of representation: not through indexical photographs but through automated data collection from a myriad of different sources constantly updated and endlessly combined to create a seamless illusion.[14]

For Valla, we are already in the Matrix. Google Earth does not look like Earth but resembles other digital materials.

Because we spend so much of our lives looking at these materials, it becomes real.

The photographer Doug Rickard similarly uses the image-stream created by Google Street View as the source for his sometimes controversial work. He bases his search for compelling images on the aesthetic of the Farm Security Administration photographs of the 1930s. Many of these photographs, like Dorothea Lange's *Migrant Mother* have become classics of American photography. To find such images, Lange, Walker Evans, Gordon Parks and others had to first go to impoverished places, then identify and capture expressive moments. Rickard did his work browsing on a computer, often deliberately visiting places that he knew FSA photographers had previously depicted. What is seen is familiar, a variant on documentary or street photography. Only the 'photographer' was never on that street and did not even take the picture.

Perhaps unintentionally, Valla and Rickard show that the two issues the society of control cannot control can be found in its online avatars as well: disaster, natural or otherwise, and inequality. Valla's images of distortions in Google Earth remind us of devastation caused by hurricanes, earthquakes and the collapse of poorly built or maintained infrastructure, which we now associate with climate-change related events. Rickard finds the lonely and disadvantaged in the supposedly 'level playing field' of the Internet. These alternative means of seeing the world – the changing natural world and social change – will be the subject of the final chapters.

The Changing World

How do we know when the climate has changed? There are problems of scale, measurement and understanding here that appear very abstract. The very concept of climate is an abstraction, a human rendering of data over time which cannot be observed in and of itself (Edwards 2010). No experiment is possible here because the scale is that of the planet itself. Scientists have devised a model called the carbon cycle to explain how the planet has sustained a favourable temperature for agriculture and settled human life over the past 12,000 years, itself a tiny measure of geological time. The carbon dioxide exhaled by animal life was exactly balanced by plant photosynthesis, while the oceans released and absorbed balanced amounts of the gas. The quantities involved are minute. The carbon balance rested at 278 parts per million in the atmosphere – a tiny percentage of invisible gas. Human activity, such as the burning of fossil fuels, has raised that number to 400 parts per million – still tiny, still invisible but now causing increasingly powerful effects in the climate worldwide. Even if all emissions were to stop tomorrow, the climate will keep changing for centuries. And yet we still can't see it, literally and metaphorically.

We have to make climate change less abstract. Here's

how it came home to me. In August 2010, I was in Guam, the US dependency in the Central Pacific, beginning research on climate change. The indigenous Chamorro people on the island have recently sought to reclaim their long-ignored rights to their country. As part of that effort, they have revived traditional navigation, sailing canoes built by hand, using no modern materials, for thousands of miles to show both that their society was not without technology and that not all technology requires environmental destruction. The navigators use their knowledge of the stars and the way land changes the direction of the ocean waves to set course.

During the conversation, a seventh-generation master navigator in this tradition, who goes only by the name Manny, explained his skill with an aura of authority. I asked if he has seen any difference as a result of climate change. He noted that he has always been able to predict the weather. His colleague explained that once a group of sailors was planning a voyage of about 2,400 kilometres (1,500 miles). Manny simply said that they needed to be back by the end of the first week in July: on 8 July that year a typhoon struck. In this equatorial region, weather patterns observed over generations have been sufficiently stable to allow for such precision, he explained. 'Now I can't tell what the weather will be,' Manny told us. That's how we see that the climate, and the world, has changed.

What we can see now is the result of human changes to the world over the long human period since the Industrial Revolution began around 1750, which is a speck in the eye of geological time. Among the most notable ongoing transformations, in addition to climate change, are the sixth mass extinction of living things, and the ever-growing clearance of over one quarter of the world's forests. Imagine a world without coral reefs, with no summer ice in the Arctic, where big animals like lions, tigers and polar bears can only be seen in zoos or carefully controlled outdoor protected areas like game parks. Welcome to 2040. The human relationship to the world is going to change fundamentally as a result of our having fundamentally changed the world. Simply put, everything will look different.

However, unless we have a highly trained eye, like Manny, these changes are not always easy to see. The most common effort to make them visible is the use of comparative formats. These can be very effective, for example the time-lapse photography used to document glacier retreats in the film *Chasing Ice* (2012). Placing twenty-five cameras in ice-fields around the world for three years, photographer James Balog created a series of time-lapse sequences at each site. Even over this short period, the resulting 'films' produced by playing the photographs continuously show a noticeable and shocking retreat of the ice. In similar fashion, James Brashears has revisited sites of famous photographs of glaciers and snow-covered mountains from the twentieth century and taken new pictures. Shown side by side, the photographs dramatically reveal the full extent to which the ice has disappeared.

Comparison has been used to depict change on the global scale. The map of the world shown in Figure 66 was produced by the British medical journal *The Lancet* in 2009 to portray the relationship between global carbon emissions and mortality. The top half of the diagram represents

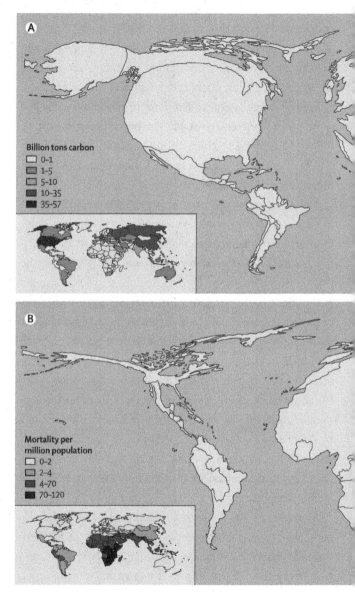

Figure 66 — *The Lancet*, map, 'Managing the Health Effects of Climate Change'

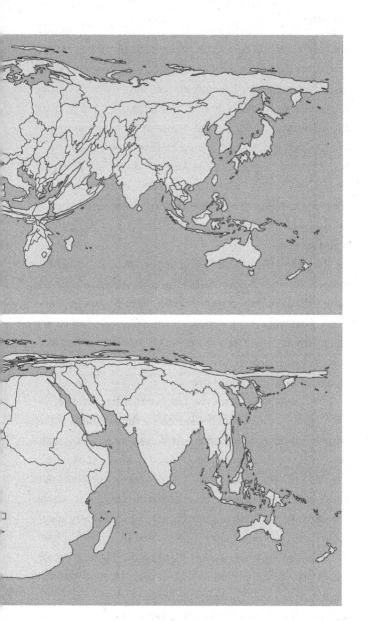

countries according to the percentage of global carbon emissions for which they are responsible. More emissions means a larger size. The bottom half shows the likely consequences in terms of human mortality (deaths, in plain English) resulting from climate change for each country. In the top half, the EU and United States are very large, clearly the greatest emitters, while Africa is almost invisible (the map would look a little different today because China has risen to the top of the emissions table). In the lower half, it is equally clear that Africa and India will suffer the greatest consequences.

The map tells us something important that we did not necessarily know before: there is an inverse relationship between the countries responsible for carbon emissions and those that suffer the consequences. Sub-Saharan Africa emits very little CO_2 but stands to lose many people as a result of climate change because of drought and other disruption to already precarious lives.

For the deniers, however, neither the diagram nor the photographs can show what is causing the warming. While 98 percent of scientists are in firm agreement that human activity is the clear cause, well-funded groups persist in calling this a debate. Many do so cynically, taking the funds that fossil fuel companies, the most profitable enterprises in the world, make available to them (Oreskes and Conway 2010). Nonetheless, another shift is concealed by all this bluster. Because climate change takes place on the planetary scale, scientists can only model its outcomes. Deniers claim to want experiments. The belief that science means observable and repeatable experiments that began with Descartes in the seventeenth century (Chapter 2) is, to this extent, over. Global understanding is, by contrast, based on computational models supported by a knowledge infrastructure: in the case of climate, these would be weather observations, satellite data, radar readings, and so on, calibrated against past measurements. It is not something that one person can do by themselves, as could the heroic scientists of the past. Knowledge itself is now a model based on an Internet network.

See change

The change revealed by these models is so thoroughgoing that geologists have named the period since the Industrial Revolution as the Anthropocene: the New Human Era. This means that we have changed the planet's fundamental geology from the rocky depths of what is called the lithosphere to the highest reaches of the atmosphere. And that means a change in the way we measure 'deep time', the way the planet's immensely long history is understood. Humans have thrived in a tiny window of geological time known as the Holocene, meaning 'entirely recent', which is a mere 12,000 years old. The Holocene is the most recent part of the Quaternary, itself a young geological era that is approximately 2.5 million years old. To put this in context, the preceding Neogene era began approximately 23 million years ago. What used to take millions of years to change now takes decades. Transformations that would have been utterly invisible to humans now take place within the short span of one person's life. We have to learn to see the Anthropocene.

For as deep time has changed, one casualty has been one of the classic ways of seeing the world. A key precept of Western thought has been to distinguish between nature, which is simply present, and culture, which is made by humans. In particular, the artist observes nature and makes it into culture – for example, a painted view of some land becomes a landscape. Now that distinction has collapsed. It has its own history that we need to trace before setting out to make the Anthropocene visible.

Since the scientific revolution of the seventeenth century, the West has prioritized what was explicitly known as 'the conquest of nature'. For the English scientist Francis Bacon, the first to call for this conquest, nature was provided by God 'for the relief of man's estate'.[1] Bacon meant that as humans are vulnerable, requiring food and shelter to survive, they can protect themselves by using the natural world as a resource. It was at just this time that Western artists began to paint landscapes, especially in the Netherlands, then a dominant economic power. Landscape was a visual representation of both the conquest of nature and the conquests of colonialism. The battle against nature was won but is now being followed by its slow collapse under the consequences of its own efforts (Nixon 2011).

We have already not only long absorbed the costs of this conflict but learned to find them beautiful. Modern beauty was often the product of climate change. In early nineteenth-century Britain, for example, newly spectacular sunsets captivated the Romantic poets. They were caused by particles of the coal being used in the new factories of the period refracting red light in the air. The Romantics used the term 'sublime' to refer to a kind of beauty that would be terrible to experience personally but was intensely moving to see depicted in art, such as a shipwreck or storm of the kind we see in the painting of the Romantic artist J. M. W. Turner.

Today's hurricanes, droughts, floods, record snowfalls and escalating temperatures create a different feeling – a constant unease as unusual weather becomes the new normal. That unease chimes with the uncanny feeling produced by the new global city, digital networks and drones. In order to have lived in a month where the world was not warming month-by-month, you need to have been born in 1985 or earlier. If you are under twenty-eight, you have never known what the pre-climate-changed world was like. Your body knows nonetheless that the drought, the floods and the rising seas are out of joint with past experience. It just feels wrong. So we have to imagine that past, 'unsee' – to use China Miéville's term – how it has taught us to see the world, and begin to imagine a different way to be with what we used to call nature. That will be seeing the Anthropocene.

The birds

One of the most visible changes to the planet has been an enormous reduction in bird populations, so central to every mythological and cultural system humans have created. Humans have been devastating bird populations for a long time, reshaping how the world looks and sounds as they go. On the Pacific island of Tonga alone, twenty-six species of birds have been documented as becoming extinct since the arrival of humans some 2,800 years ago. Modern seafaring dramatically increased the pace of such extinction. Dutch sailors arrived on the island of Mauritius in 1598, where they found edible, flightless birds they called the dodo. Sailors and other travellers ate the huge birds in considerable numbers, while the pigs and macaques introduced by the travellers feasted on their eggs. The last accepted sighting of the dodo was in 1662, one of the first casualties of the conquest of nature.

The transformation is astonishing seen in the longer time-frames of extinction by means of natural selection. The 'background' extinction rate (meaning the number of extinctions that would occur in the absence of human intervention) is very low. It would take 400 years for a single species of bird to become extinct without human involvement. The dodo has been a fixture in popular culture since the nineteenth century because it was the first modern sign that humans could change change itself.

The death of birds became a scientific curiosity across early modern Europe that stood in for the conquest of nature as a whole. The painter Joseph Wright of Derby depicted such a death in his *An Experiment on a Bird in the Air Pump* (1768). The air pump was invented in 1659 by the scientist Robert Boyle, who used it to demonstrate many of the otherwise invisible properties of air. His experiment, in which a bird was placed in the pump while the air was removed, demonstrated the necessity of air for life because the bird died. A century later, the experiment was as much entertainment as science, performed by self-described 'natural philosophers' in lecture halls and private houses. (These are the kind of experiments that climate deniers today imagine

Figure 67 — Green (after Wright), *An Experiment on a Bird in the Air Pump*

being performed to test climate change.) Such is the scene dramatically depicted by Wright. The bird flutters in the vacuum pump, to the distress of the children, while adults conduct learned conversation on the spectacle. Science is represented as reason triumphing over sentiment, gendered as masculine and feminine respectively. The dramatic candlelight and the biblical appearance of the experimenter add to the tension of the scene. Most actual experiments used small native birds like larks and sparrows, but Wright painted a cockatoo. These tropical birds were newly known to English people thanks to the voyages of Captain Cook. The few specimens in the country were rare and expensive, so it would have been an unlikely candidate for death by scientific amusement. By painting a tropical bird, Wright intended to highlight and visualize the symmetry between the conquest of nature and the conquest of new territory overseas that British philosophers had been making throughout the lifetime of the air pump (see Chapter 3).

Such experiments were not controversial because modern Westerners considered birds as inexhaustible resources. One of the most dramatic examples of this misconception is the Passenger Pigeon. These birds were so populous in North America that it challenged belief. The famous ornithologist and bird artist John James Audubon was so amazed by the 'countless multitudes' of birds that he attempted to estimate them as they flew past him in Kentucky in 1813. He calculated that there were no fewer than 1,115,136,000 pigeons in the single 'flock' that he saw. He was equally stunned by the visual beauty of it all:

> I cannot describe to you the extreme beauty of their
> aerial evolutions, when a Hawk chanced to press upon
> the rear of a flock. At once, like a torrent, and with a noise
> like thunder, they rushed into a compact mass, pressing
> upon each other towards the centre. In these almost solid
> masses, they darted forward in undulating and angular
> lines, descended and swept close over the earth with inconceivable velocity, mounted perpendicularly so as to resemble a vast column, and, when high, were seen wheeling and
> twisting within their continued lines, which then resembled
> the coils of a gigantic serpent.[2]

As they flew by, however, humans waited with guns. Audubon described how at each location the birds passed, people would shoot as many as they possibly could, both

Figure 68 – Audubon, *Passenger Pigeons*

to feed themselves and to fatten domestic pigs. The birds sold at market for a penny apiece. While Audubon feared that humans might extinguish the birds, he could not bring himself to believe that it could happen. A century after he encountered the pigeons in Kentucky, the last known Passenger Pigeon died in a zoo in Cincinnati, Ohio, on 1 September 1914, just as humans set about slaughtering themselves in the First World War.

Ironically, Audubon's drawing of the birds has now become a memorial to this extinct species. It shows two birds 'billing', a courtship ritual in which one bird feeds the other. The female bird, above, nurtures the more brightly coloured male bird, for as Audubon noted, 'the tenderness and affection displayed by these birds toward their mates, are in the highest degree striking'. For a century, it has not been possible to experience this relationship, once an integral feature of North American life, let alone see the spectacular manoeuvres used to deter predators by the massive flocks of birds. Like all Audubon's drawings, it was itself made using corpses, rather than drawn from live birds. Audubon created a device using wires to pin the bird into the position he wished to draw it, as you can see from the forced position of the lower bird's wings to display the colourful tail.

His classic *Birds of America* (1827–38) is filled with accounts of shooting birds and otherwise obtaining dead birds from the flourishing bird markets that he visited, from New York to New Orleans. In his own time, this process was unremarkable but today it seems to describe an everyday theatre of cruelty that led to extinction. While people shoot birds less than they did, birds continue to decline as humans increase their settlements and the climate changes.

In 1962, the science writer Rachel Carson changed the way people understood the environment with her book *Silent Spring*, first published in the *New Yorker* magazine

(Carson 1962). Carson showed that the pesticide DDT was causing tremendous harm to people and animals, including birds. The title of her book came from her effort to imagine a spring without birdsong, caused by the damage DDT does to the shells of birds' eggs. The combination of this powerful image and her convincing evidence resulted first in restrictions on DDT use and then its ban altogether. We might wonder if Americans today, living in cities behind double-glazing and plugged into their headphones, would be so moved by a threat to birdsong.

The 2007 Audubon Society report on the citizen-count of the twenty most common birds in America found that since 1967 the average population of the common birds in steepest decline has fallen by 68 percent; some individual species nose-dived as much as 80 percent.[3] A follow-up study in 2014 suggested that climate change threatens half of the bird species in America. John James Audubon's nineteenth-century volumes now document a cluster of extinct species and many more in sharp decline. We live on, and look at, a different, emptier, less song-filled planet than he did.

Modern beauty

Work that intended to capture the new or the everyday is now also a monument to environmental destruction or climate change. Just as natural history drawing captured soon-to-be extinct animals by accident, so too did the painting of the new phenomena of modern industrial life highlight the process of climate change without the artists realizing what was really happening. This 'double play' capacity can be seen across modern Western art. The city has become the habitat for the majority and we have naturalized it in art, photography and film. We can learn to look again at these works to see how humans have changed the world, and then we could develop ways of seeing the planet which might be part of the solution. To do so, however, we have to 'unsee' the ways in which we have come to see this change as beauty.

The effects have been dramatic since the beginning. There is no modern painting more widely reproduced and taught than Claude Monet's *Impression: Sun Rising* (1873).

Without diminishing our appreciation of Monet's handling of colour and light, I want to stress that this is a painting that at once reveals and makes beautiful human environmental destruction. Coming late to the Industrial Revolution, France was just experiencing the smog produced

Figure 69 — Monet, *Impression: Sun Rising*

by industrial coal use in the mid-nineteenth century. The port of Le Havre in Normandy, seen in Monet's picture, was well known for its smokiness. The effect was featured in a range of French visual culture such as popular photographs, postcards and paintings from the middle of the nineteenth century on. Monet grew up in Le Havre, which was the main French port for transatlantic passenger shipping, predominantly steamship traffic. In his painting, traditional rowing boats can be seen isolated in the foreground. In the background, industrial machinery dominates the scene, such as the angular cranes to the right. Coal smoke pours forth from the chimneys of three steamers clearly depicted in the left middle-ground of the painting. Overall, the painting generates the vivid set of sense impressions that gave first this work, and then an entire movement, its name. Coal smoke is yellow, the yellow that predominates at the top of the painting. In the early morning, the time depicted in the painting, the smoke encounters both blue morning light and the red of the rising sun, producing the array of refracted colour that makes Monet's painting so stunning in the original.

There is a good deal of artfulness in Monet's apparently spontaneous effort to seize the moment. The mix of light and smoke combines to form what we might call a very modern form of beauty. The steamers can barely be distinguished in the smog, giving them the appearance of factory chimneys. Somehow they fight their way out of the water, like a latter-day Leviathan, the legendary sea monster, adopted by philosopher Thomas Hobbes as the symbol of the state. The steamers are literally and metaphorically the source of power. The painting is made as if looking from an unusually high viewpoint. Perhaps Monet was looking out of a high window or from the rigging of a ship. Whether he physically put himself in such a place is beside the point. Monet gave visual form to the conquest of nature, transforming the once-fearsome ocean into a domesticated, human dominated object. It's as if the function of the sea was now to be looked at.

Here the humans who have made the world in their own image look at their creation and see that it is good. While Monet's contemporaries at first experienced his work as shockingly modern and new, it soon became comfortably familiar, as it remains today. The painting made the transformation of the world by modern industrial processes not only visible but beautiful. Beauty has no practical purpose. So although the actual smoke was the sign of industrial work, the painting emphasizes its hand-made qualities. Monet wanted us to realize that his art was not simply a copy that a factory might make. It is for and about the leisure class, not the working classes. In reality, the smog was a dangerous by-product. The modern idea of beauty transformed the sensing of the colour and smell of coal smoke into an indication of the continuing conquest of nature.

Two years later, Monet fully realized this world-view in his small but dense painting *Unloading Coal*. A fleet of coal barges from the mines in the north of France enter the picture space from the bottom left to the centre, almost as if invading it. Coal, itself the product of a very demanding form of manual labour, is carried off the barges for use in this industrial suburb of Paris. The workers cannot be distinguished individually, precisely because as individuals they

Figure 70 — Monet, *Unloading Coal*

do not matter. What counts is just the unloading of the coal. Like mining, it's back-breaking work. From here, the coal is transported by means we cannot see to factories like the one in the background, once again pouring out smoke. Those factories produced goods like the iron for the modern bridge and the commodities being transported across it by carts. A gaslight, the visible sign of modern human dominance over nature, can just be made out on the left of the bridge. The bridge appears to be a visibly 'higher' level of existence, one dominated by manufactured goods, and artificially lit. The figures are no more distinct or individualized up there. Some are at rest, or watching the underlings carry coal, so you would rather be one of them. The key spaces of modern industrial society – production and consumption – are linked into one visualized system here.

Just as with *Impression* a few years earlier, *Unloading Coal* is constructed from an unusual mid-air viewpoint, perhaps the view from a train window as it crossed over the river heading to Paris. As we saw in Chapter 4, the moving image as seen from the train is often taken to be the precursor of cinema. Monet here made the moving modern world into a still. This freeze-frame accounts for the strong sense of movement in the painting, given coherence by its overall warm tone, that subdued yellow hue, which is the product of coal smoke. The degradation of the air is again seen as natural, right and by extension beautiful. The changed world is now so built in to our senses that it determines our very perceptions, and so it has become beautiful and aesthetic.

If beauty is what is known as the aesthetic, art here produces a sensory anaesthetic to the actual physical conditions.[4] Whereas watching coal being unloaded on a smoggy day might not be an elevating experience, looking at Monet's painting of such a scene is exactly that. Just as nineteenth-century art had pictured storms and mountains as beautiful rather than threatening, Monet changed our perception of the modern city. Indeed, the invention of medical anaesthesia in the nineteenth century was one of the most dramatic reductions in human suffering ever known, so the dulling of the senses was not always perceived as a bad thing. A remarkable example of how this sensory anaesthesia actually worked in practice comes from New York. If we look at George Wesley Bellows' classic painting *Forty-Two*

Figure 71 — Bellows, *Forty-Two Kids*

Kids (1907), we see a group of naked children getting ready to swim in the East River on a hot day.

The assumption is that they are poor, from the Lower East Side of the city, where immigrants were then congregated in large numbers and dreadful conditions. The water is black. It was not a metaphor. At that time, all the bodily waste of the 6 million people living around New York Harbor was piped straight into the water. You could also find many dead animals in the river, not to mention industrial waste. In the nineteenth century, oyster beds were so flourishing in New York that they were one of the key food sources for the city. By the early twentieth century, they had all died.

Officials charged with dealing with the disposal of sewage could not understand why there was no public outcry or even perception of the waste. In 1912, five years after Bellows made his painting, a British scientist commented after a tour of the harbour: 'I am surprised that a city claiming to be one of the first in the world should allow such a disgraceful condition of affairs to exist.' City sewage officials noted in amazement:

> The people of New York seem strangely indifferent to the polluted condition of the harbor. They have recently built some of the finest and most expensive hospitals and apartment houses on the shores of the most polluted large part of the inner harbor, namely, the Upper East River, where that fetid stream is joined by the black and malodorous Harlem.[5]

The point was that, while the 'great unwashed' working classes might have been expected to be willing to live with dirt and smells, so too were New York's elites. Even today, any rainstorm that generates over half an inch of water flushes raw sewage into New York's rivers. Swimmers and surfers know to stay out of the water at local beaches the next day. The desire to live in the modern city was so great that it anaesthetized the senses, or at least allowed people to disregard what they saw and smelled in the water. The image of the city replaced its material reality and became a new reality.

Such selective perception was by no means unique to New York. For over a century, London was afflicted with dense smogs produced by burning coal. Known as 'peasoupers', and often referred to incorrectly as fog, this persistent smog became a feature of London life. Tourists expected the fog, and Londoners missed it when away. It became a character in nineteenth-century fiction, such as the famous opening to Charles Dickens's *Bleak House* (1852–3):

> Fog everywhere. Fog up the river, where it flows among green aits and meadows; fog down the river, where it rolls defiled among the tiers of shipping and the waterside pollutions of a great (and dirty) city. Fog on the Essex marshes, fog on the Kentish heights. Fog creeping into the cabooses of collier-brigs; fog lying out on the yards and hovering in the rigging of great ships; fog drooping on the gunwales of barges and small boats. Fog in the eyes and throats of ancient Greenwich pensioners, wheezing by the firesides of their wards; fog in the stem and bowl of the afternoon pipe of the wrathful skipper, down in his close cabin; fog cruelly pinching the toes and fingers of his shivering little 'prentice boy on deck.

It is as if the fog (really the coal smogs) is now the background to all natural and human activity. The smog is so naturalized that it can itself even be polluted by the dirt of the city. Yet it makes the shops turn on their lights two hours early and dims out the gaslights of the streets. It is everywhere, and 'at the very heart of the fog', says Dickens, is the High Court of Chancery. This court dealt with cases concerning property. The fog symbolizes the dominant place of the rule of law in modern life. It reaches every corner of our lives, every object that surrounds us. In the eyes of imperial culture, law separated the 'civilized' from the 'savage', the result of the conquest of nature. Fog was the visible by-product and symbol of that conquest. *Bleak House* was not so sure, for the interminable case of *Jarndyce v. Jarndyce* at its centre destroyed the lives of all it touched.

Exactly a century after the publication of *Bleak House*, the Great Fog (as it was then known) of December 1952 brought twilight at noon to London. Photographs show dim outlines of landmarks perceived through the haze. A later study estimated that it killed some 12,000 people by exacerbating lung disease and other breathing difficulties, more than four times the casualties of 9/11. But if you look in newspapers, diaries and other sources for those days, it's hardly mentioned. *The Times* newspaper noted that fog held up traffic and mentioned breathing difficulties only for cattle at the Earl's Court market. After a century, fog just came with London, in the way that smog (by then properly named at least) was later associated with Los Angeles. In retrospect, the Great Fog is often associated with the passing of the Clean Air Act of 1956 that finally led to the end of the peasoupers, if not of the smog. In fact, the Act followed a Private Member's Bill, showing that there was no great urgency from the official point of view.

Olympic coal and steel

In his classic description of urban poverty in the Great Depression of the 1930s, *The Road to Wigan Pier*, George Orwell described how

> Our civilization . . . is founded on coal, more completely than one realizes until one stops to think about it. The machines that keep us alive, and the machines that make machines, are all directly or indirectly dependent upon coal.[6]

In many ways, it would be fair to say that it still is, especially if steel is added to the mix. Records show an acceleration of environmental destruction, especially carbon emissions, since the beginning of globalization in the 1980s. Despite its high levels of carbon emissions, 30 percent of UK energy is still derived from coal, while 39 percent of US electricity came from coal in 2013. China, which is now the leading user of energy worldwide, derived 69 percent of it from coal in 2011. Coal has again become the single largest source of global carbon emissions, overtaking those from vehicles. When the world is changing, it makes little sense to measure results nation by nation. We have to think in terms of cause and effect planet-wide, meaning we have to relearn how to see the world as a whole. We cannot see from the partial viewpoint of our own nation or region, but need to bring together different points of view so as to see the Anthropocene. Perhaps we are still anaesthetized by our pleasure in modern urban living, unseeing what its costs are at home and elsewhere.

Looking at the dynamics within Monet's *Unloading Coal* allowed us to see the formation of the Anthropocene in Europe. Today we need a global version of this way of seeing. Rather than a single frame, we should think of connecting a series of such frames into a 'film' that would allow us to see the structures, networks, histories and effects of the Anthropocene. This film would not be the view from the train so much as the view from the ground. It is the view first of all of the landless. In Brazil, one percent of the population owns 45 percent of the land. Worldwide, 20 percent of the hungry are landless food producers, while there were

2.2 billion people earning $2 a day or less in 2011, according to the World Bank.[7] It is also the view from a sustainable farm, from the bottom of a skyscraper or from the informal housing around global cities.

Sustainable and fairly owned land is the opposite to the global city using coal for energy and steel for building construction. Coal and steel production links mining nations such as Australia, Brazil, India and South Africa to China and the developed-world economies in a network of mining, production and final use of energy and construction in global cities. This is the network we need to understand in order to think visually and see the changing world.

When the Olympic Games were held in Beijing in 2008, there was tremendous concern in Western media and sports circles about the effects of air pollution. Some athletes arrived wearing facemasks. It has recently been estimated that 750,000 people a year die from pollution-caused illness in China. In the winter of 2012–13, 600 million Chinese people lived under a cloud of smog that covered 1.3 million square kilometres (about 500,000 square miles); It was visible from space, disappearing only occasionally.[8]

By contrast, when the first day of the 2012 London Olympics coincided with an air-quality alert, no one mentioned it. The story about London's air quality was supposed to be a good one. Since 1990, UK emissions have been reduced by about 21 percent, mostly due to declining coal use.[9] But the picture is complicated by carbon emissions due to imported goods and services. The UK emissions from imports increased 23 percent from 1997 to 2004 but have since fallen due to the financial crisis.[10] London will not meet European Union air-quality standards before 2030, it has since been announced. Chinese smog is very visible in Western media, whereas first-world failings are ignored.

How can we start to see past the haze? Let's pursue this Olympic parallel through the coal and steel networks. Both the Beijing and London Olympics engaged top artists to create steel monuments as a central part of their projects. In Beijing, Ai Wei Wei helped design the dramatic Bird's Nest Stadium (officially known as the National Stadium), while Anish Kapoor created a giant public sculpture known as *Orbit* for the 2012 Games. Both the stadium and the sculpture were made from steel. The global steel industry has been booming in parallel with the rise of China. According to 2012 figures, China produces over 700 million tons of steel a year, half of all global production, compared to just 88 million tons in the United States.[11] Steel production is a spectacular source of carbon emissions. According to the Organization of Economically Developed Countries:

> Steel production accounts for 5% (8% including power, mining and ferro-alloys) of global CO_2 emissions. The steel industry is the largest industrial CO_2 emitter (30%).[12]

The smog around the stadiums was, then, directly connected to the steel with which they and the monuments around them are made.

The Bird's Nest Stadium alone used 110,000 tons of steel. The remarkable elliptical shape of the Stadium was at first designed to support a retractable roof. When this was cancelled because of cost, the result was a beautifully

Figure 72 — Beijing National Stadium

original form. The apparently porous building was designed by Swiss firm Herzog & de Meuron working with Chinese architect Li Xinggang. It was inspired by Chinese ceramics and cost $300 million (about £190 million). As Li said, 'In China, a bird's nest is very expensive, something you eat on special occasions.'

The Stadium adopted the concept of nature as a resource for human need and turned it into a monument to Chinese progress. Lit up by fireworks for the opening and closing ceremonies, the Stadium was indeed a stunning sight. Even more remarkable was the invisible transformation produced by Chinese officials during the Games. By dint of compulsory reduction of industrial activity and keeping cars off the roads, China produced clean-air days for the Games. Not only that, scientists later calculated that the reduction in emissions was 0.25 percent of the entire global target to keep under a 2° Celsius rise in average temperature. The surprising lesson is that, were other global cities to follow

Figure 73 — Kapoor, *Orbit*

this lead, it would actually be possible even at this late stage to contain global warming.

Four years later in London, the Games were marked by a specific steel monument, Anish Kapoor's *Orbit*. In this instance, the steel came from ArcelorMittal (after which the sculpture is now named), a global steel corporation run by

Lakshmi Mittal, Britain's wealthiest man. His company had revenues of over $94 billion in 2011 and outlets in 60 countries. Throughout the Olympic year 2012, there was passionate industrial action over job losses at an ArcelorMittal plant in north-eastern France, which became an issue in that country's Presidential election. These issues were invisible in London, where most of the discussion was about *Orbit* as an artwork. It was very different from Kapoor's best-known works that tend to be smooth, curved, sometimes highly reflective forms. *Orbit* is a visually confusing tangle. Kapoor spoke of 'making something that was continually in movement'.[13]

From what seems to be the intended viewing point, a long extension heads from the bottom left into space on the right, distracting and breaking the flow of the piece. It looks better from the other side. Even so, what is this? Kapoor's goal of giving visual form to 'instability' has perhaps been realized too well. There's a viewing platform on top of what looks like one of those terrifying circular exits to European car parks. The piece is, simply, the spectacle. It works as long as you look at the Olympic Stadium next door, smaller than I expected. Or look towards the increasingly dramatic skyline of the City of London, including Renzo Piano's dramatic 308-metre (1,000-foot) high 87-storey skyscraper The Shard, completed just in time for the Olympics, capped with a 500-ton steel spire. Another way of looking at the City would be to see it as home to many of the scandals associated with the 2008 financial crisis, such as the manipulation of the LIBOR interest rate. Look the other way from *Orbit*, though, and the view is of Stratford, an as-yet-ungentrified part of the East End, dominated by unlovely tower blocks and a tangle of roads, overhead power lines and railway tracks.

In interviews, Kapoor claimed an affinity with Vladimir Tatlin's legendary *Monument to the Third International* and the Eiffel Tower. Gustave Eiffel's steel pyramid was built for the International Exhibition of 1889 in Paris. These were the mass spectacle tourist events of their time, displaying products from around the imperial world. In fact, many of the pavilions even included residents of the countries they represented as human displays. The International Exhibition was an unabashed celebration of the conquest of nature and the rise of Western 'civilization' in its place.

By contrast, although it was never built, Tatlin's spiralling Constructivist tower was intended as a homage to the 1917 Russian Revolution. His design (1919–20) aspired to outreach the Eiffel Tower and thus symbolize the dominance of Communism. For Lenin, communism was famously 'electricity plus Soviets', embodying just as much determination to conquer nature. If we imagine Kapoor's sculpture as the successor to both empire and Communism, it might be seen as the 'Monument to Globalization'. It makes sense in this context, four years into the financial crisis that began in 2008, that it has the strong feeling of instability. From this perspective, you can imagine that what *Orbit* actually looks like is a folded-in combination of the characters for pounds, dollars and euros: £ / $ / €. In that way, it really was the most appropriate monument that there could have been.

Visual thinking for the Anthropocene era

If there are to be new ways of imagining ourselves in the world, there will need to be a new visual way of thinking for the Anthropocene era, perhaps even a Monument for the Anthropocene (one is actually being planned by the Argentinian artist Tomás Saraceno in Toulouse, France). A good place to start would be the documents produced for over twenty years by Canadian photographer Edward Burtynsky of what he calls the 'manufactured landscape' created by mining. Such landscapes can be seen all over the world and are entirely artificial, the building blocks of the Anthropocene. In his 1985 photograph of the Westar Open Pit Coal Mine, Burtynsky captured the sheer scale of the mine at Sparwood. Trucks and other equipment in the middle ground are dwarfed by the concentric circles formed by coal removal that laid bare the hillside, forming a new apocalyptic human-generated vista. We might call this anthropocene landscape. Located close to the Banff National Park in British Columbia, Sparwood is a small community otherwise famed only for displaying what it claims to be the world's largest truck. Good roads connect the mine to the Interstate system so that the coal can be distributed quickly. The coal from the open mine is now used in the manufacture of steel. In 2014, thirty years after the picture was taken, the mine was expected to be active for another twenty-nine years. These interconnections and networks are part of what made the Bird's Nest Stadium and the *Orbit* possible but remain unseen, the material side to globalization that most prefer to ignore, just as the anthropocene landscape is experienced only by those who have to work there.

Another vital part of this thinking is to show how colonial histories continue to shape energy production. The artist Sammy Baloji (b. 1978) makes us see how the colonial history of the Democratic Republic of Congo's second largest city, Lubumbashi, in Katanga province, is directly connected to the global city and its digital networks. Born in Lubumbashi, Baloji trained in the DRC and in France. His extensive and widely exhibited *Mémoire* (2006) project consists of montages in which Africans and Europeans photographed in the colonial era appear in front of present-day mine works. Once again, the sense of haunting and the uncanny reappears in our global present.

Made in very large-scale format, Baloji's montage of black-and-white and colour film is visually arresting. His work attests to the long history of exploitation in the region.

Figure 74 — Baloji, *Mémoire*

Beginning in the 1920s, Belgian colonists began to mine the immense copper deposits in Katanga, often using forced labour, especially during the Second World War when demand for copper was high. After independence, the Gécamines state-owned mining company became equally famed for its output and its corruption in the 1980s. Following subsequent decades of war, the mines photographed by Baloji are now post-industrial ruins. Broken-down buildings and enormous deposits of trailings (what remains after the ore has been extracted) create an apocalyptic landscape. Baloji describes how

> my current works have a direct connection with the colonial past, which gave birth to the cities of Katanga province. These cities were built upon mines. The latter belong to Katanga's history. The essence of my question lies in the daily life of Congolese people.[14]

In 2006, Congo was at the bottom of the United Nations Human Development Index and close to bottom of its transparency index, indicating high levels of corruption. As many as half a million people, including many children, now work as subsistence miners in the region, meaning that they dig ore-bearing rock out of the ground and sell it individually. The demand for copper comes largely from China, which takes up to 40 percent of the global supply in order to make consumer goods such as computers, refrigerators, cars and plumbing equipment. In the United States, the Dodd-Frank Act (2010)[15] now forbids the use of so-called 'conflict minerals', meaning minerals produced under duress or in war conditions. Despite efforts by companies like Intel, the global flows of the mineral market make it very hard to know whether the computer I am using to write this chapter has Congo copper in it or not.

What techniques should we use to make such global flows apparent? *Coal + Ice*, a 2011 international collaborative documentary exhibition drawing on the work of more than thirty photographers, set out to make visible the connections between increased coal use and melting ice fields.[16] In the words of its curators Jeroen de Vries and the photographer Susan Meiselas, *Coal + Ice* 'visually narrates the hidden chain of actions triggered by mankind's use of coal. This photographic arc moves from deep within the coal mines to the glaciers of the greater Himalaya where greenhouse gases are warming the high altitude climate' (*Coal + Ice* 2010). The installation asked spectators to see the connections between modernity and climate change without pushing easy answers or dictating conclusions.

Figure 75 — Installation view of *Coal + Ice*, Yixian, China

James Brashears' photographs of retreating glaciers hang here above powerful documentary photographs of the labour of Chinese miners. The story is not simple. Mining communities are close knit, generating not just financial benefits but pride and solidarity. At the same time, the work is difficult and dangerous, with planetary consequences. Ending mining would have benefits for the climate but would damage these human communities. It is up to the viewer how to frame the comparison. It takes time and installations like this allow us to do the visual thinking necessary to imagine these histories and to begin to devise alternatives.

We should develop our skills in this kind of visual thinking to understand human interaction with key natural systems, such as rivers, which are now undergoing dramatic change. Consider the vital Mississippi River. Long the means of transporting America's wealth from the days of the Cotton Kingdom in the south to today's grain shipments going downriver, passing oil tankers heading north, the Mississippi is a key national artery. It both waters – and now increasingly floods – many states. In 1944, Harold Fisk of the Army Corps of Engineers, which is responsible for federal waterways, made a remarkable map of the Mississippi River flood plain.

Fisk's large-scale and monumental map shows a swirling set of meanders and bows formed over the long expanse of geological time. It makes deep time visible. The 'modern' course of the river is depicted in white at the centre of the tangled weave of its former trajectories. The result looks more like a William Blake painting than a geological diagram. The two hundred years in which European-Americans have been trying to modify the course of the river are too brief to

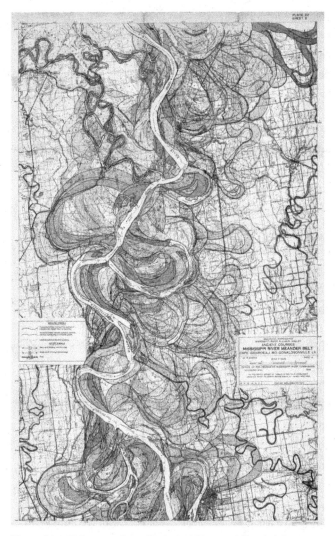

Figure 76 – Fisk, map of the Mississippi River and flood plain

visualize on such a scale. The map gives us a sense of the river as a living thing, with a history and memory, rather than an inanimate flow of water. It creates a sense of comparison and history in one frame and shows that any attempt to confine the river to a single course is likely to be futile.

By contrast, twenty-first-century maps of the river by the Corps show it as a straight line, constrained between impassable levees that are only as strong as their weakest point. The city of New Orleans discovered this to its cost when the levees broke after Hurricane Katrina in 2005. You might not at first even recognize this as a map of a river.

The Corps' map visualizes the conquest of nature. It turns the whorls and swirls of the river that Fisk had mapped into a set of straight lines and data points. That river does not exist. By the same token, nor can the Corps successfully contain it within those boundaries.

The Army Corps of Engineers is a group of soldiers involved in maintaining and extending the conquest of nature. The Corps refers to river water as the 'enemy' and adopts what has been called a 'fortress' model for the preservation of cities. It has helped eliminate wetlands and bayous which provided some natural protection against flooding. Since Katrina, most calls for the restoration of New Orleans have followed that fortress or 'hard' model. After Hurricane Sandy hit New York City (2012), the preferred term there has been 'resilience', meaning sea walls and other barriers. The alternative is 'soft development'. Soft development emphasizes the restoration of wetlands, swamps, shellfish beds and other means of absorbing or diverting floodwaters. It allows rivers to flow more naturally. Compare Fisk's map of

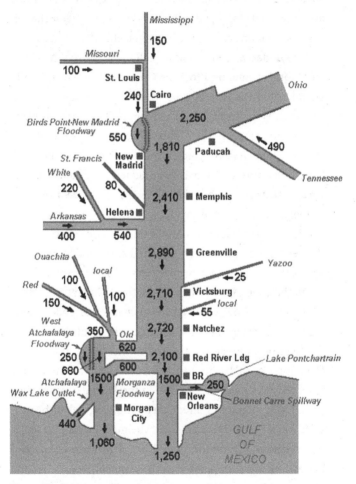

Figure 77 — US Army Corps of Engineers, diagrammatic map of the Mississippi River

the Mississippi to that made by the Corps and we can see how the Army try to replace curves with straight lines. In what remains a very militarized society, the 'hard' options are more culturally and politically palatable, even though the 'soft' options are more likely to be effective.

The underlying question is really how we see the changing world around us. From the Greek philosopher Aristotle we inherited the idea of unity of time and place, meaning that what is depicted should be seen from a particular place over no longer than a single day. The visual system of perspective, known to the Ancients and revived to dramatic effect in the European Renaissance, added the injunction that what is seen should be taken in by a single spectator from a single, identifiable place. To see the changing world, we will have to set aside all of these time-honoured strategies. We need to compare across time and space and learn to see from other people's perspectives as well as our own.

As the examples in this chapter from Guam to the Mississippi River show, we also have to change our understanding of time. Deep time is changing in front of our eyes. If we don't take into account the worldwide situation, we will constantly be taken by surprise. Developed nations largely ignored reports of sea-level rise in the Pacific because they assumed it would not affect them and were taken by surprise when the 2011 tsunami overwhelmed sea walls in Japan and released substantial quantities of radiation from the Fukushima nuclear plant into the atmosphere, the Pacific Ocean and beyond. 'No man is an island,' wrote John Donne in seventeenth-century London. We are now all connected and change itself is changing.

Changing the World

On 1 January 1994, as the world was getting over its New Year's Eve festivities, the Zapatista rebel army came out of the jungle in Chiapas, Mexico, and declared 'Ya Basta!' (Enough!). The action was timed to coincide with the commencement of the North American Free Trade Agreement (NAFTA) that removed trade barriers between Mexico, the United States and Canada. The EZLN (Zapatista Army of National Liberation / Ejército Zapatista de Liberación Nacional) was formed to create alternatives to globalization for the local Maya and other groups, concentrating on civil rather than armed resistance. The Zapatistas made skilful use of media to spread their concept of a politics 'from below, for below', issuing a series of 'Declarations from the Lacandon Jungle' online. They saw changing media and politics as two parts of the same process. The Zapatistas have a talent for media-friendly events. Their spokesperson, Subcomandante Marcos, became something of a media personality, always appearing wearing a ski mask and smoking a pipe.

Many have credited the Zapatistas with the invention of 'hactivism', activism online that seeks to disrupt the operations of government or corporate websites. One of the first such actions followed violent incursions by the Mexican army, leading to the displacement of 5,000 people. The Zapatistas and their allies then undertook a virtual sit-in on the Mexican government's website on 18 June 1999. Due to the relatively unsophisticated web security at that time, people were able to participate in the virtual sit-in using simple HTML script. They insist that these actions are electronic non-violent civil disobedience, rather than criminal actions. In their Sixth Declaration, the Zapatistas announced that they envisage 'a world where there is room for many worlds, a world that can be one and diverse'.[1]

Taken together, these approaches amounted to a new form of 'representation' for the era of globalization. 'Represent' has two distinct meanings here. First, representation is the way we depict events and experience in other form, whether on film, in photography or any other medium. For the Zapatistas, participatory media events like the virtual sit-in are not just a form of publicity for their cause, but rather are examples of the kind of world they hope to create. With the spread of global digital culture, this participatory approach to media is far more widespread and understood than it was in 1994. Second, 'represent' means the representative system of government, in which individuals are elected or appointed to represent the interests of others. However, once in place, these representatives have a good deal of latitude to decide how to act. The Zapatistas wanted to empower people to govern themselves in what has been called 'direct democracy'. Although people found them inspirational worldwide, it was only in Chiapas that they were able to create lasting change. In fact, their concept of participatory democracy using digital media was better suited to the new global cities. Working through this doubled concept of representation is the second component of the new visual thinking required by the era of globalization.

Rebel cities

Global cities around the world, from Cairo to Istanbul, New York to Madrid, have indeed since become places of protest, claiming what the scholar David Harvey has called 'the right to the city' (2013). It is here that the young, urban, networked majority are questioning both forms of representation. In 2001 Argentinian protestors brought down no fewer than five governments using the slogan: 'They do not represent us'.[2] Their call raises the question as to whether the new global majority can represent itself both politically and visually, or whether the visible oligarchies generated by globalization will continue.

The double question of representation first jumped from peasant areas to global cities in Argentina. After its military dictatorship collapsed in 1983, Argentina took on extensive loans from the International Monetary Fund (IMF). In the 1990s, it was compelled by the IMF to introduce a stringent austerity regime. Even these measures failed, so that in November 2001, the government converted people's personal money into an asset that could be used to repay the international loans. The result was that if you went to the ATM, you could not get money out. Financial order had broken down. On 19–20 December 2001, the people of Buenos Aires spontaneously revolted, followed by the rest of the country. A city of some 13 million, Buenos Aires extends over roughly 200 square kilometres (80 square miles). Its people forced both the existing government and then four new governments to resign over the space of a month (Sitrin 2006). This was the first of the new attempts to change the world by making the global city into a rebel city. The new majority had found a new way to call for change. The exasperated call 'Enough!' had given rise to 'Que se vayan todos!' (Let's get rid of all of them!) because they do not represent us. Which means that 'we' have to do it ourselves.

The movement for self-representation that began in Latin America spread worldwide, enabled by social media and other Internet-based platforms. It gained global attention with the Arab Spring and the subsequent global Occupy movement in 2011. These movements tried to find new means to represent people, who participated both as individuals and as 'the people'. Using social media and political action, the people first claimed a name, whether as the *indignados* in Spain, the 99% in the United States, or simply 'the people' in Tunisia and Egypt. Then they found a space: Tahrir Square in Egypt, Puerta del Sol in Madrid, Zuccotti Park in New York. From these places, the movements claimed not to represent but to *be* the *indignados*, the 99% or the people respectively. They asserted the right to look and be seen, online and in city space. This new self-representation used smart phones, graffiti, websites, social media, demonstrations and occupations.

In Egypt, those in the square claimed to *be* Egypt, not to represent it. This claim was sufficiently powerful that leaders fell and regimes changed. For a time, it seemed as if the

Arab Spring and other movements really might change the world. The rebel cities rarely dominated the entire country, however, so national leaders were able to reassert their claim to be the true representatives of the people as a whole, often making strategic use of state media. Now that this wave of claims to the right to the city has subsided, we can look back and see how it changed visual culture, from North Africa to North America. These movements were the first to use global social media to try and create visual thinking about representation and social change: where do we go from here?

2011 and after: North Africa

If visual culture is a performance, as we saw in Chapter 1, then the Arab Spring began with a dramatic opening act. In Tunisia, one person became the representative of the consequences of state repression. Tarek al-Tayeb Mohamed Bouazizi was a fruit-seller who dramatically set himself on fire on 17 December 2010 in frustrated protest against police interference with his work and the regime in general. Such self-burning had precedents in the 1960s, as part of protests against the Vietnam War and the Soviet occupation of Eastern Europe. Bouazizi lived in a small town called Sidi Bouzid, close to a mining town that had been occupied by protestors for six months in 2008.

One year earlier in the city of Monastir, 200 kilometres away from Sidi Bouzid, a young man selling doughnuts had also had enough of the police restrictions on his attempts to trade and set himself on fire in front of a state building. Nothing happened. A year later, Mohamed Bouazizi repeated the act, whether in conscious imitation or not, and Tunisia underwent a revolution. Why did his act resonate with the public when the earlier one did not? The difference was simple: the diffusion of the news by Facebook and other forms of peer-to-peer communication. Facebook did not cause the revolution, but it allowed for the dissemination of information. People were ready to act because climate change had brought drought, leading to high food prices, in the context of political corruption, mass unemployment and widespread unrest.

Social media enabled people to set aside the unseeing of this crisis required by the regime. People still had to act as a result of that information for there to be social change. After Bouazizi's death, further acts of protest were extensively covered by al-Jazeera, the Qatari satellite news channel. When these new forms of representation interacted, his suicide set in motion a process whereby people concluded enough was enough, the regime did not represent them, and thus they had to take over their own country.

During the 2011 revolution, the French self-styled 'artivist' (artist/activist) who calls himself JR realized a transformation of Tunisian visual culture was happening. The only portrait photograph seen in public in Tunisia for the past forty years had been that of the former Tunisian dictator Zine al-Abidine Ben Ali. JR organized an 'inside out' representation of Tunisia, putting the formerly invisible people into public space. It was intended to be a transformation of unseeing into seeing. Working in conjunction with Tunisian bloggers, and using only local interlocutors and photographers, the goal was to create a series of one hundred portraits of people who had participated in the revolution. Printed as 90 × 120 centimetre posters, they were fly-posted across four cities in Tunisia, including startling examples stuck on the former secret-police commissariat and on the façade of one of Ben Ali's former houses.

The project was called 'artocracy', meaning the rule of art. Yet this open-access project was strongly criticized in Tunisia. 'Why only one hundred?' was the common refrain. For the revolution was widely held to have been the work of the people, not a sub-set of heroes. JR's posters did not – and perhaps could not – adequately represent the revolution. No one wanted to replace autocracy, the rule of one person, with the artocracy of one hundred, even as a joke. JR's visual thinking was not sufficiently sharp.

The Egyptian uprising of 2011 and after was perhaps the most striking effort to recreate representation and transform visual thinking. The young population of Egypt's cities rose up in the face of a significant food crisis caused by the North African drought, which was surely linked to climate change. As the Center for Climate Security has shown, while they did not 'cause' the events, 'the consequences of climate change are stressors that can ignite a volatile mix of underlying causes that erupt into revolution'.[3] Social media networking was also an essential factor, as was the density of urban population in Cairo and Alexandria, where over ten million people live. Fully 70 percent of the Egyptian population is under the age of thirty, while official youth unemployment was 25 percent or more during the revolution. All these factors combined to make the Egyptian revolution. It would be better to say that 'Egypt' was remade in Tahrir

Square and it continued to be redefined and re-represented until the Army put a stop to it.

Although it is a major global corporation, Facebook was a key tool of the 2011 uprising in Egypt. It began with 400,000 'likes' on the 'We Are All Khaled Said' Facebook page in 2011. Khaled Said was a blogger, who was arrested and tortured to death by Egyptian police in 2010. The virtual assembly on his Facebook page enabled in part the mass demonstrations of January 2011 by making visible a substantial alternative social group. For over thirty years, the dictatorship of Hosni Mubarak had been able to prevent any such demonstrations in physical space. Facebook was also used to communicate news and dates for action. The date 25 January 2011 was set on Facebook as the day for mass action, a call disseminated both in the streets and via Twitter. Over 90,000 people 'liked' the Facebook page. Hundreds of thousands actually took to the streets, for social media had catalysed the movement in ways that had not been seen before. The sheer size of the turnout took everyone by surprise, from the organizers to the police and the outside world. The regime no longer represented the people.

A street battle on 28 January opened Tahrir Square to the newly configured people. It was perhaps the closest to the nineteenth-century revolutions that we are likely to witness. The difference was that it was seen live via the Internet and on al-Jazeera, whose coverage of the Arab Spring circumvented the efforts of the dictators to downplay what was happening both within their own countries and globally. The battle of Kasr Al Nil Bridge, which decided whether or not the demonstrators would take over Tahrir Square, was a real and violent conflict. Nonetheless, perhaps because it was being watched live, the Egyptian police did not use live fire and the Army pointedly did not intervene. This was not a contest of sheer power, as a military battle might be. It was a battle over who controls public space: the ordinary people or the forces of public order? After 29 January 2011, the people were able to create their own order for the first time in thirty years of dictatorship.

Tahrir Square was an unlikely space for liberation, an irregular shape formed between sets of government buildings, and usually a jammed traffic intersection. It owed its existence to the British colonial administration wanting to clear lines of fire in front of administrative buildings, as we saw in nineteenth-century Paris (Chapter 5). Tahrir became a space of active resistance to the dictatorship from below, as well as the place to provide the protesters with health care, food and media 'broadcast' via the Internet. In this way, the square became a kind of technology in itself. It created the very possibility of political action and gave new meaning to the concept of public space. In short, it was both a new form of visual representation and the claim to be politically representative, cross-hatched in a new experience of space. The square became, as it were, a projector, superimposing these new ideas over the old Tahrir Square, where the secret police and Mubarak's New Democratic Party had both had their headquarters.

From its first days, the revolution was condensed into the slogan: 'The people want the fall of the regime'. In this phrase, a self-image formed where there was none before. In the eighteenth century, the philosopher Jean-Jacques

Rousseau described what he called 'the general will', the force and power of public opinion.[4] This had been repressed in Egypt for thirty years. Over the course of eighteen extraordinary days, a new general will was formed, and it was captured live on TV. Its subject, the people, forced the dictator to yield because it was clear that they, not he, represented Egypt.

After the occupation of Tahrir Square, the social movements in Egypt produced new forms of visual thinking, including 'street art', graffiti and video collectives. Graffiti was startlingly new for Egyptians, because the dictatorship had maintained absolute control of public space. Graffiti is a way to reclaim public space for discussion. It can reach people who might not see mainstream media, let alone go to an art gallery, in a country where UNICEF documents 26 percent of the adult population as being illiterate, disproportionately women.[5] As such, the graffiti that flourished in Cairo and Alexandria until the coup led by General Sisi in June 2013 could drive political debate and open new senses of possibility. Particular spaces, like Mahmoun Street in Cairo, where Mubarak's Ministry of the Interior had once orchestrated a reign of police terror, became central locations for this visual discussion. After the coup, a law was proposed against 'abusive graffiti' which appears now to be in effect.

One young graffiti artist, Mohamed Fahmy, who calls himself Ganzeer (meaning 'bicycle chain' in Arabic), describes himself as a 'contingency artist'. In other words, his work responds to the needs of the moment in whatever way seems right. He thinks of this as participatory art in this sense: 'art that participates in dealing with the immediate struggles and concerns of the audience'. It thinks along with its audience, rather than for them. While he accepts that the revolution itself has failed, at least for now, 'it does not mean that the effects of the revolution should not find their way into art and culture'.[6] His first work during the revolution included a graffiti memorial to a sixteen-year-old protestor killed by police and a widely used PDF pamphlet on how to conduct a protest. He later described how he was motivated to write the handout when he saw that the demonstrators could not respond to police tactics.[7] The pamphlet gave specific ideas on how to organize, and advised people that a photocopy is more secure than a web post which can be tracked down by the authorities. After the fall of Mubarak, Ganzeer set himself the marathon project of creating street portraits of all 847 people who died in the revolution, known as 'the martyrs'. He had only accomplished three of these portraits by the time he left the country in 2014, making it unlikely that his martyrology will ever be accomplished.

The Supreme Council of the Armed Forces, who were running Egypt in 2011, persisted in painting over these three memorials. Ganzeer responded with a piece showing an enormous, full-size Army tank bearing down on a man on a bicycle carrying a tray of bread. The human figure was dwarfed by the machine. As the Arabic for bread also means 'life', the graffiti piece suggests that Army rule was being opposed to free life. In the photograph in Figure 78, you can also see the signature work of Sad Panda on the far right, added, as it were, to Ganzeer's piece, as well as posters and other graffiti. The wall has become a place for visual conversation.

Figure 78 — Ganzeer, *Tank and Bread*, Cairo

In May 2011, activists held a Mad Graffiti Weekend to re-store the memorials. Ganzeer circulated a sticker called 'The Freedom Mask', showing a masked, gagged head with the caption: 'Greetings from the Supreme Council of the Armed Forces to the beloved people. Now available in the market for an unlimited time'. This sarcasm led to his arrest but after he tweeted his situation, so many people came to the police station that he was released without charge.

In October 2012, Ganzeer held a formal exhibition of his work, called 'The Virus Is Spreading', at the Safarkhan Gallery in Cairo.[8] The work explored questions of freedom, sexual identity, censorship and Islam. The pieces included a dramatic image of a blindfolded man sewing his mouth shut. A wounded cat, who has lost one eye and is covered in bandages, was an updated version of the symbol of Egypt. The show extended across several floors, making beautiful use of calligraphy on the walls, following the widespread use of such graffiti in Cairo. One example from early 2013 post-ed to a blog read: 'O stupid regime understand my demand:

freedom, freedom'. Other graffiti artists like Sad Panda also participated. The exhibition was quickly denounced by Islamists as heresy. Ganzeer published an open letter in reply:

> Do you know any group of liberals that have prevented the construction of a mosque? Has a liberal person ever criticized an art exhibition on the grounds that it was Islamic – and even sought punishment for its participants?[9]

His point was that no liberal Egyptian had sought to prevent Islamic groups from carrying out their way of life but the reverse was certainly not true. In response to this active threat of censorship, Ganzeer and his fellow street artists like Keizer and Sad Panda used the Internet to archive their work (cairostreetart.com). A Google Maps mash-up indicated where and when the work was posted. Users were invited to 'like' the link on Twitpic and Flickr but not on Facebook, which was now too carefully under surveillance. The website cairostreetart.com was taken down in 2014, presumably by agents of the new military regime or by people not wishing to endanger themselves.[10] In April 2014, Sampsa, a street artist also known as the 'Finnish Banksy', told a reporter:

> In Egypt, some of these street artists are being visited by police weekly. They . . . are trying to find more subversive ways because they are being tracked down, via social media, physically. Thousands of people have been picked up and put in jail.[11]

Others have turned up drowned in the Nile. Ganzeer himself had to leave Egypt in May 2014 after being accused of activism for the Muslim Brotherhood, and the country is currently undergoing a new wave of repression in which social media is closely monitored.

For the online counter-archive was a key tactic in creating new means of engagement. An important example is the non-profit media collective Mosireen. In their own words, Mosireen was

> born out of the explosion of citizen media and cultural activism in Egypt during the revolution. Armed with mobile phones and cameras, thousands upon thousands of citizens kept the balance of truth in their country by recording events as they happened in front of them, wrong-footing censorship and empowering the voice of a street-level perspective.[12]

In January 2011, at the height of the uprising, Mosireen was the most-watched not-for-profit YouTube channel in the world. Their activities came to centre on documentary and media activism, meaning using video to show the world what was happening in Egypt in the face of domestic censorship and international ignorance. They archived the revolution, with 10 terabytes of video already collected by 2013. Mosireen hosted busy open workshops in media techniques on a pay-as-you-can basis. Screenings of their videos were arranged outside, circumventing the need for Internet access. During the occupation of Tahrir, they even created Tahrir Cinema, projecting films for the occupiers.

The heart of Mosireen's work is their videos. These remarkable films, from the centre of the revolution, were made with skill and courage, often beautifully edited and composed. They were posted online within days of the events they depict, and English subtitles appeared on a second version hours later. The films from the January revolution of 2011 are extraordinary documents from the middle of a popular uprising. They consist of montaged video clips taken on the streets of Cairo during the protests. There is no commentary or narrative. Contrary to many Western media reports, women are involved in all the actions. Detail is provided by comments from the participants themselves. In a film like *Martyrs of the Revolution* (2011), Western viewers may be shocked by the graphic violence.[13] Military police vehicles are seen running over protestors in the street at high speed. A soldier deposits the body of a protestor in the trash. Wounds from rubber bullets, truncheons and live ammunition are shown. The cameras are right among the crowd, so the filmmakers placed themselves at the same risk as the other participants. A clip shows a young man installing graffiti, another shows a woman declaring that everyone in Tahrir is now her son Ahmed, killed in the protests; still another shows a defenceless young man on his own being shot by the authorities. At the end, a three-column list of the names of those killed takes over two minutes to play.

Perhaps the most polished of Mosireen's films during the revolution appeared in January 2012. *The People Demand the Fall of the Regime* took as its title the signature chant of the Tahrir movement.[14] It montaged scenes of everyday life in Cairo with the preparations for a protest, culminating in a vast crowd chanting the words of the title. The film is set to Wagner's overture to *Das Rheingold*, a famously beautiful and haunting piece of music. A woman cradles a newborn, a singer composes a new piece and people celebrate in front of a public television. But this is not a romantic view of the revolution. Again, viewers are confronted with the brute force used by Mubarak's regime in its desperate attempt to cling to power. The final shot shows the still astonishing sight of Tahrir under occupation, with its cluster of tents in the centre where food and medical care were distributed, surrounded by people as far as the eye can see.

In 2013, the film *The Square*, nominated for an Academy Award for Best Documentary, gave a vivid insight into the daily lives of five activists in the Egyptian revolution, including key organizers with Mosireen. There are extensive disagreements over both tactics and fundamental principles in the group, but they find themselves drawn together by their resistance first to the dictatorship and then to the regimes that came after. The film follows their transition from believing that overturning the dictator Mubarak would amount to a revolution in itself, to a realization that the process may be long and that their task is to oppose each and any regime that does not enable the representation of the people in all senses. We see young people being trained in how to use cameras and take video. At one point, Ahmed, a young street organizer, reflects on the relationship of the two forms of representation. 'As long as there's a camera, the revolution will continue,' Ahmed suggests. Meaning, that as long as people can see what is being done, they will continue to demand a regime that truly represents them. Ahmed sees the overthrow of the Muslim Brotherhood and the subsequent takeover by General Sisi and the army as just two more steps

As long as there's a camera, the revolution will continue.

Figure 79 — Still from the documentary *The Square*

in this process. While it may seem from outside that he is far too sanguine, only time will tell.

The North African uprisings of 2011 combined social media with street protests and online archiving to create a new form of visual culture activism. In highly censored societies like Tunisia and Egypt, the chance to depict yourself and others in public, let alone to express political opinions, was a rupture with decades of past experience. The resulting visual thought created hope, made the revolutions possible and helped drive them forward. This is not to call the Arab Spring a set of Facebook revolutions. But it is the case that a networked and young population, experiencing food scarcity due to climate change, used visual activism on- and offline as a key component of their urban uprisings.

2011: Occupy Wall Street

Perhaps it is not surprising that this visual activism and visual thinking was strongly resonant in New York, a hub both for professional media and for aspiring media producers. In July 2011, the Canadian magazine *Adbusters* launched a call to 'Occupy Wall Street'. *Adbusters* began as a not-for-profit movement to repurpose advertisements, using skilled designers to create different meanings than were intended. Known as 'culture jamming', this kind of satirical play on mass media is intended to cause the viewer to question what he or she sees.

One well-known example is this fake ad, which shows the golden arches of McDonalds displayed on a heart monitor in a hospital emergency room. Adbusters placed them above

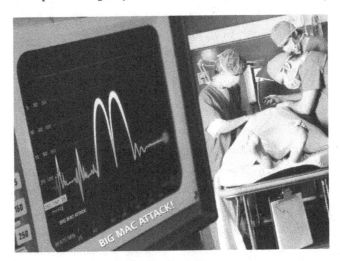

Figure 80 — Adbusters, 'Big Mac Attack'

the caption 'Big Mac Attack!' to remind us that eating fast food is quite likely to give us a heart attack from all the fat and salt in the burgers. On 13 July 2011, Adbusters posted this call on its blog:

> On September 17, we want to see 20,000 people flood into lower Manhattan, set up tents, kitchens, peaceful barricades and occupy Wall Street for a few months. Once there, we shall incessantly repeat one simple demand in a plurality of voices.
>
> Tahrir succeeded in large part because the people of Egypt made a straightforward ultimatum – that Mubarak must go – over and over again until they won. Following this model, what is our equally uncomplicated demand?[15]

Six weeks of organizing later, on 17 September, about two thousand people gathered in a little-known public-private park, called Zuccotti Park, close to Wall Street.

Almost at once, Occupy Wall Street (OWS) departed from the Adbusters plan. Adbusters had hoped that street

Figure 81 – Adbusters, 'Occupy Wall Street'

protestors could take over Wall Street itself, one of the most heavily policed streets in the world. Knowing this, New York organizers planned to camp close to Wall Street, rather than on it. Nor did OWS want to push one single demand as Adbusters had suggested because it did not want to claim to be representative. Rather, each of its ideas – and many more – became a slogan on someone's sign. OWS refused to make demands on the grounds that this was an autonomous self-governing movement. Unlike Tahrir, which claimed to be Egypt as such, OWS declared New York City to be 'occupied' for two months (17 September–13 November 2011). This was not a military occupation but a takeover of the city by the people who are normally overlooked on Wall Street, such as the young, the unemployed and the homeless.

OWS seemed to give a place and voice to such people, especially those who had been made invisible by financial globalization. Collectively, they did not claim to be or to represent all the people but rather the 99%. The one percent left out was the group identified by economist Joseph Stiglitz as the wealthiest of all:

> The upper 1 percent of Americans are now taking in nearly a quarter of the nation's income every year. In terms of wealth rather than income, the top 1 percent control 40 percent ... Twenty-five years ago, the corresponding figures were 12 percent and 33 percent.[16]

From Occupy's point of view, the one percent continued to benefit despite creating the recession that first ruined the global economy and then they were bailed out for doing so. In discussions at Zuccotti Park, activists decided that if the wealthiest were the one percent, then everyone else must be the 99% (Graeber 2013). Perhaps the most significant effect of the Occupy movement was to reintroduce discussions of inequality into mainstream American culture, although little has been done to change the wealth divide.

A Tumblr called 'WeAreThe99%' was created to allow people to tell their stories, and to represent themselves. Tumblr is an easy-to-use blogging site that does not require its users to have their own website or host. It is mostly used by young people as a kind of digital scrapbook.

Figure 82 — Still from 'WeAreThe99%', Tumblr

'WeAreThe99%' was a creative and emotionally powerful form of visual thought that appealed first to young people but then became widely known. People posted photographs of themselves holding written texts describing their situation, a politicized form of the 'selfie' we looked at in Chapter 1.

Often they concentrated on the way in which they had tried to play by the rules but ended up in financial disaster anyway. Student debt was a key topic, as were unemployment, lay-offs, cut-backs, benefit 'retrenchments', outsourcing, pension depletion – all the vocabulary of everyday financialization. The stories were all the more powerful and moving because they were compressed into one image with only as many words as could fit onto a single sheet of paper.

The signature object of the Tumblr, the hand-written sign, migrated to the actual Occupy sites. Made with a felt-tip pen on bits of cardboard box, such signs conveyed the force of authenticity that the mass-produced signs often seen at professionally organized protests do not. Many were simply spontaneous. There was wit, irony and insight in a way that the canned slogans of organized campaigns often fail to achieve.

Figure 83 — 'Occupy Wall Street' sign

The sign in Figure 83 expressed the sense that making an alternative autonomous society was going to be a messy process and that there would be a period of transition. And that it would be fun.

Whether protestors were aware of it or not, the British artist Gillian Wearing had used this tactic in her conceptual art project 'Signs that Say What You Want Them to Say and Not Signs that Say What Someone Else Wants You to Say' (1992–3). Wearing gave a piece of paper and a marker to people for them to write whatever they wanted and then photographed them.

In one example, taken from a series of about 600 altogether, the sign read: 'I signed on and they would not give me nothing', meaning that a claim for unemployment benefit had been denied. The author appears to be homeless, given the location in a Tube station, his rucksack behind him and the can of lager next to it. That might not be the case, but when we see these pictures in an art gallery, we cannot follow up with questions as to why the claim was denied or what the person is doing. We don't even know his name. Perhaps that is part of the reason that Ganzeer said that conceptual art was useless for the Egyptian revolution.

By contrast, signs at OWS were always an invitation to a conversation. Many people would stand with their signs on the edge of Zuccotti Park facing Broadway, precisely in order to get into conversation with passers-by. Popular signs might win honks from passing cars or waves from tourist buses. Others provoked intense debate.

There was no more popular sign with the occupiers than this: 'Shit Is Fucked Up and Bullshit.' The sign scatologically

Figure 84 — 'Occupy Wall Street' sign

and humorously summarized the entire WeAreThe99% Tumblr and indeed the whole movement. It bluntly stated how everyday life was often experienced during the recession in a way that the mainstream media would never allow. Its phrasing was far more common online, especially on Twitter. Both media scholar McKenzie Wark and philosopher Simon Critchley used the slogan as the title for their writing about Occupy. The anonymous author of the iconic sign clearly had art skills. The sign was made from canvas stretched over a frame, like a painting, and the artist's calligraphy followed the rules of graphic design. OWS had many professional artists associated with it, who formed working groups with names like Arts and Labor. These signs were not shown in a gallery, like Wearing's prize-winning work, but were being looked at by people who may not even have been aware of such spaces. It was a reclaiming and a reimagining of representation, the assertion that there was something to see there.

Spray here for social movement

At first, OWS had relatively small impact. It was the distribution of photographs and video on social media showing police violence that proved decisive in making the occupation take off. A curious meme (a widely reproduced and circulated visual image) of women being pepper-sprayed by police has since emerged across three continents as a catalyst for social movements. Pepper spray is an intensely concentrated form of chilli pepper made into an aerosol. Patented in 1973, it was first used by New York police in 1994.[17]

Figure 85 — Still from YouTube video of Officer Bologna pepper-spraying protestors

On 24 September 2011, thousands of people watched a video of the pepper-spraying of three young women by a New York Police Department officer. The women were already 'kettled' behind unbreakable plastic netting and posed no apparent threat to public order. The sight of young white women being subjected to this disproportionate force was shocking, though it is of course true that people of colour have long suffered such violence at the hands of the police, only mostly out of sight and offline. The video went viral. Soon versions were posted in slow motion with captions inserted to show what was happening, keeping the event in the news. The hacker collective Anonymous then identified the officer involved as Deputy Inspector Anthony Bologna on 26 September by enlarging a still of his badge and using its information to find him.

Two days later, 28 September, a second video emerged showing Bologna spraying other protestors, apparently just to move them out of his way. Although police officials defended him at the time, they declined to support him when the case came to trial a year later. During the Civil Rights era, television had exposed the violence of police. There were no TV cameras at Occupy for the most part and Bologna, an older officer, might not have been aware of how new media could make worldwide news. At this point, the occupation was days old and only a few hundred were involved. On 5 October, the New York area trade unions called a solidarity march for OWS. Fifteen thousand took part. There were actions at university campuses across the country. Officer Bologna deserves part of the credit.

In June 2013, although global media had declared the wave of social justice uprisings to be finished, it began again. When a new movement appeared in Istanbul, Turkey, in defence of Gezi Park, the last public green space in Istanbul. The Erdogan government wanted to turn the park into a mall-cum-Ottoman-theme-park-cum-mosque. Like the reconstructed Berlin castle discussed in Chapter 5, this re-building was intended to complete a new form of social authority in the city. The heavy-handed police response again created a popular backlash.

Figure 86 — Orsal photograph, *Woman in Red*

Once again, a photograph of a woman being pepper-sprayed went viral. Now known as *Woman in Red*, the photograph shows a gas-masked policeman spraying a well-dressed and unthreatening young woman with such force that her hair flies in every direction. This photograph was, however, taken and distributed by the giant media agency Reuters, rather than by a protestor. What had previously happened from within the social movement was now being orchestrated by the mainstream media.

And then it happened again in Brazil. Protests against a rise in basic transport costs, while the government was building sports stadia for the World Cup and the Olympics, suddenly brought tens of thousands on to the streets. A sense of fairness and social justice was mobilized. And this time it was an Associated Press photographer who got the image of a woman being pepper-sprayed.

As in Istanbul, a woman carrying nothing more than a summer bag is directly targeted by a fully armoured policeman. Such use of pepper spray has become routine and the only difference is that the mainstream media are now reporting it. When they do, it has given a noticeable impulse to the social movements concerned. The concept 'Move on, there's nothing to see here' is unpacked in this action. On the one hand, the police make sure protestors can see nothing by dint of spraying their eyes with pepper. But the media representation of the scene brings many other people into the protest. What began as a social media meme has become a mainstream media pattern of reporting that unintentionally reinforces the events that are being covered. This set of effects, from protest to social media, mainstream media and back to protest, is indicative both of how the new global situation has changed and how change itself is now a key subject for anyone interested in the visual.

Occupy and other urban protests had the advantage of working in a media environment that was already saturated with images and therefore with an audience skilled in visual analysis (even if they might not put it like that themselves). The 99% Tumblr and the pepper-spray video went viral in part because their audience were adept at sharing and disseminating media content. There is an emerging method from such protests. To create a meme takes conscious effort. The 99% slogan was the result of intensive discussion among highly aware activists, and is usually credited to David Graeber, now a Professor at the London School of Economics. For the meme to work, it requires a pre-existing network. If something is shared among a few dozen people, it's unlikely to gain traction. Reach a thousand people and the friends of their friends on Facebook alone will total over 25 million. OWS built a network where it could directly contact hundreds of thousands of people, so, at two degrees of separation, enormous audiences could be reached.

None of that matters unless the 'performance' is right. The very word 'occupy' struck a chord in 2011, creating offshoots that ranged from Occupy Museums to Occupy Technology and Occupy Student Debt. It has continued its global impact in 2014 with the emergence of Hong Kong's remarkable Occupy Central With Love and Peace movement. The handmade and homemade quality of the media objects created by Occupy Wall Street resonated still further in a moment where focus-group tested, professionally produced work often seemed wildly out of touch with the difficulties of everyday life. Put in a formula, visual culture activism in 2011 involved creating, performing and disseminating memes in urban public space and across social media networks to involve, extend and create a political subject, such as the WeAreThe99%.

Perhaps this wave of revolt in global cities has ended or, as its protagonists might put it, been repressed. Nonetheless, uncoordinated and often very distinct moments of unrest remain a feature of the global scene. The Kiev Maidan protest of 2013–14 was a determined revolt against the government whose motives were never clear and whose consequences led to the division of Ukraine. Protests in Bangkok, Thailand, at the same period were actually opposed to democratic elections. The Hong Kong actions in 2014 were directed in theory at the 2017 Chief Executive election, but also had in mind the long-term future of the former colony and its anticipated return to the Chinese system in 2046. After police shootings went unpunished across the United States in 2014, the #BlackLivesMatter movement disrupted and reclaimed public space across the country. So the motivation and outcome of the urban revolts are not consistent or a given.

If we look worldwide we can see that the combination of increased urbanization, mass youth unemployment and climate-change-driven unrest is set to continue. In 2013, youth unemployment in Greece and Spain exceeded 50 percent, while in the Eurozone countries as a whole it was 25 percent. In South Africa, 52 percent of young people were unemployed, a figure that reached 59 percent in Detroit, Michigan. According to the Palestinian Authority, 41 percent of young Palestinians were without work in 2013 and significant unrest followed in 2014, especially in Gaza.[18] A 2014 report on the national security consequences of climate change, issued by a team of eleven retired generals and admirals, noted that these impacts had gone from being a threat multiplier to 'catalysts for conflict' in the course of seven years from 2007.[19] In the same week, scientists reported that the West Antarctic ice sheet was set to break off and melt, causing in and of itself a 3-metre (10-foot) rise in sea level, albeit possibly over centuries.

The urban population, mostly living in delta and coastal cities threatened by sea-level rise, continues to grow, with over one billion people now living in informal housing (Davis 2006). That means one in three people in the developing world live in slums. All the new urban experience takes place in less than 3 percent of the world's surface, providing the intensity of contact to create change. And these people are increasingly networked, even in developing nations. The selfie was just the first form to emerge from this networking. Imagine a selfie of 'the people'. It's hard to see all this as anything other than a catalyst for continued change in ways that we cannot yet fully anticipate or imagine. In all the reports on these changes, one of the most notable features is the call for a better imagining of the future. At the heart of imagination is the image. Visual culture has to respond day to day in its effort to understand change in a world too enormous to see but vital to imagine. At one level, it serves as a form of academic 'first responder' connecting present-day situations to longer histories. It seeks to understand the total visual noise all around us every day as the new everyday condition. And it learns how to learn about how the visual imagination, visual thought and visualizing combine to make worlds that we live in and seek to change.

Visual
Activism

So what then is visual culture now? It has evolved into a form of practice that might be called visual thinking. Visual thinking is something we do not simply study; we have to engage with it ourselves. What we might call visual culture practice has gone through several versions in the past twenty-five years and has now converged around visual activism. For many artists, academics and others who see themselves as visual activists, visual culture is a way to create forms of change. If we review the interpretations of visual culture outlined in this book, we can see how this concept has emerged.

When visual culture became a keyword and focus of study in and around 1990, as we saw in the Introduction, it centred on the question of visual and media representation, especially in mass and popular culture. The shorthand for understanding the issues concerning visual culture at that time was to say it was about the Barbie doll, the *Star Trek* series and everything concerning Madonna. By which we should understand that people were centrally concerned with how identity, especially gender and sexual identity, was represented in popular culture, and the ways in which artists and filmmakers responded to those representations. I do not mean to say that these issues no longer matter but that the ways in which we engage with them have changed.

The South African photographer Zanele Muholi (b. 1972) is one key example. She calls herself a 'black lesbian' and a 'visual activist'. Her self-portrait resonates with Samuel Fosso's, which we looked at in Chapter 1 (with Figure 19). Both use leopard-print as a sign for 'Africa'. Although both are wearing glasses, Muholi's heavy frames suggest she is an intellectual, while Fosso's sunglasses were part of his parody. Muholi's hat places her in modern, urban South Africa. Above all, her direct look at the camera claims the right to see and be seen.

Her work makes visible the tension between the freedoms offered by the South African constitution and the realities of homophobic violence encountered by LGBTQI (lesbian, gay, bisexual, transgender, queer or questioning and intersex) people every day. Legal protection for people of all sexual orientations exists in theory but it is ineffective day-to-day in the townships. Muholi's work shows how she and other queer South Africans are engaged with their lives and loves in the face of this violence (Lloyd 2014). She wishes to be seen as a black lesbian and to be accepted as such by her peers. In 2014, Muholi gave the keynote speech at the International Association of Visual Culture conference in San Francisco, itself titled 'Visual Activism'. For the hundreds in attendance, the questions implied by her work were global: What does it mean to be seen to be a citizen in a global era? Who represents us at local and national levels in a globalized society? If the state cannot

Figure 87 —Muholi, *Self-Portrait*

back up its own declarations with actions, how do we represent ourselves, visually and politically?

These questions resonate with the shift in thinking
through representation that began around 2001 with the
participatory movement slogan 'They do not represent us',

which we discussed in Chapter 7. The notion that 'they do not represent us' now appears more like a recurrent theme in modern history, from the Chartist claim to represent England to the Arab Spring.

The financial crash of 2007 and onwards in Ireland led to unemployment, emigration and a widespread sense of crisis in government. Art and museums have become a place to try and think through how to respond to this crisis. Artists Megs Morley and Tom Flanagan came across some notes made in 1867 by Karl Marx for a speech on Ireland which seemed uncannily familiar:

> The situation of the *mass of the people* has deteriorated, and their state is verging to a crisis [similar to that of the 1846 Famine]. (Marx 1867)

Morley and Flanagan asked three writers to imagine their own speeches on 'The Question of Ireland'. They then had actors perform the speeches, which they filmed in Ireland's National Theatre of Irish language, the Taibhearc.

Figure 88 — Still from Morley and Flanagan, *The Question of Ireland*

The result was a three-screen, hour-long film that combined the visual language of avant-garde cinema with the classic political rhetoric of the popular speech. It is a real performance but now it seeks to find a national rather than a personal identity. Morley and Flanagan go back to the revolutionary past to look for possible futures. The second segment (in the still above) meditates on how Ireland was created as a new nation less than a century ago, with great hopes, but it has not been able to realize them. The speaker concludes that what is needed is a revolution, but not in the classical Marxist sense: 'Revolutions are about vision . . . a revolution of vision, of purpose, maybe hope'. This revolution is not imagined as violent or confrontational but begins with the simple act of 'loving ourselves' in a country known for the self-deprecatory wise-crack. Although this was a film shown in art galleries and museums, its hope and intent was to create change in Ireland, above all a change of vision.

For what has become clear is that the implication of 'they do not represent us' (in all the senses of that term) is that we must find ways to represent ourselves. Visual activism, from the selfie to the projection of a new concept of the 'people', and the necessity of seeing the Anthropocene, is now engaged in trying to make that change. That effort takes place against the backdrop of ongoing war, from Afghanistan to Ukraine, and especially across the Middle East. It is not a short-term project but one that involves considering how we live our lives as a whole.

In Detroit, the 99-year-old activist and philosopher Grace Lee Boggs begins every meeting with a question: 'What time is it on the clock of the world?' In the opening shot of the film *American Revolutionary* (2014, director Grace Lee), Boggs muses: 'I feel so sorry for people not living in Detroit.' As you watch the (then) ninety-five-year-old carefully wheel her walker among one of the many urban ruins of the city,

Figure 89 – Still from Grace Lee, *American Revolutionary*

you may wonder if she can be serious. Boggs has devoted her life to Detroit. She moved there in 1955 when it was the global hub of the automotive industry. Detroit gave the world the assembly line, affordable transport, personal consumer credit to buy cars – and, as Boggs likes to point out, global warming via the automobile. In her view, we have now to engage in what she calls 'visionary organizing' to think about how life after industrial, fossil-fuel-based culture might be possible. She sees this as exciting and liberating, a chance to move 'beyond making a living to make a life'. Despite the poverty in the city – now officially affecting 42 percent of the 81 percent African-American population – Grace Lee Boggs sees the future as beginning again in Detroit.

In Grace's view, we all live in some form of 'Detroit'. What is called 'globalization' is a transition from the industrial economy to something else. What was created at the Ford factories in Detroit was the assembly-line system of production. A worker carried out the same task over and over again because this division of labour enabled the factory as a whole to produce more cars. Most of the work in a modern Ford factory is done by robots, welding and painting in showers of sparks that might be dangerous to people. One of the tasks of the remaining human labour force is to think of ways to make the process still more efficient. A group of Toyota workers realized that their paintshop could reduce its staff from eight people to three if some changes were introduced. Toyota rewarded these individuals but dismissed five out of eight employees in their paintshops worldwide. Not without reason, the Italian philosopher Paolo Virno has called the new way of working 'Toyota-ism', just as the assembly line was known as 'Fordism' (2004).

Visionary organizing is a way of thinking about how we might use our creative energies to better ends than cutting jobs and increasing profits. It is another form of visual activism. People around the world are coming to similar conclusions and finding new ways to engage with how to imagine change. In Germany, an opinion poll found that 24 percent of young people expressed the desire to become an artist. I don't think that suddenly a quarter of all Germans want to be painters or sculptors. Rather, art might seem to be the only way to live a life for yourself in the global economy, as opposed to the dominant so-called 'service economy' in which we work, not for each other but for someone else's profit. This desire to live otherwise lies behind the world-wide surge in participatory media, from YouTube channels to Snapchatting, and performance. Teen bloggers and video-channels on YouTube are finding audiences in the tens of millions, while 32 million watched the 2014 League of Legends videogame championships in South Korea. Even museums are becoming involved. The proposed M+ museum is described as a 'new museum for visual culture in Hong Kong'. Scheduled to open in 2018, it has already provoked a lively debate in the city as to what visual culture means: is it a way of thinking about contemporary art in the global city? Or is it a set of everyday practices such as graffiti, calligraphy, martial arts films and other aspects of Hong Kong's dynamic city life? Even the most traditional of museums are changing. In 2014, London's Victoria and Albert Museum held an exhibition called *Disobedient Objects* that set out to show 'how political activism drives a wealth of design ingenuity and collective creativity that defy standard definitions of art and design' (vam.ac.uk). One example was a giant inflatable cobblestone, created by the Eclectic Electric Collective for use in street demonstrations. The balloons were a pun on the cobblestones formerly used to build barricades. They make fun of the militarized way that governments try to control their citizens when police in riot gear have to run around trying to pop them. The two moments suddenly interacted with the appearance of Occupy Central. Hong Kong activists downloaded instructions on how to make a gas mask from the Victoria and Albert Museum website, while an Occupy Umbrella – the symbol of the Hong Kong movement – quickly found its way into the London exhibition.

Another side of the same situation was seen in Ferguson, Missouri, after the police shooting of Michael Brown on 9 August 2014. Acting on the understanding that Brown had raised his hands, activists created the meme 'Hands Up,

Don't Shoot' within days. Whereas most memes are thought out and planned, this was a spontaneous re-enactment of what were held to be Michael Brown's last words. The meme became known instantly through livestream and social media. 'Hands Up, Don't Shoot' is one of the first products of the interaction of the Snapchat/Selfie generation, with direct action in the streets, because it creates a new self-image of the protestor. It makes visible what was done even though it was perpetrated out of sight of any media depiction or representation. The grand jury decision not to indict Officer Darren Wilson, the policeman who shot Brown, for any crime, took the 'Hands Up' meme across the United States and indeed the world, with solidarity actions in London and elsewhere, using the slogan.

In visual activist projects, there is an alternative visual vocabulary emerging. It is collective and collaborative, containing archiving, networking, researching, and mapping among other tools, all in the service of a vision of making change. These are the goals that the tools of visual culture, which I set out in the introduction, seek to achieve. In 1990, we could use visual culture to criticize and counter the way that we were depicted in art, film and mass media. Today, we can actively use visual culture to create new self-images, new ways to see and be seen, and new ways to see the world. That is visual activism. At the end of this book we can perhaps put it still more simply.

Visual activism is the interaction of pixels and actions to make change. Pixels are the visible result of everything produced by a computer, from words created by a word-processor to all forms of image, sound and video. Actions

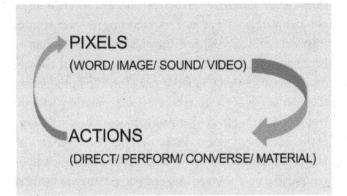

Figure 90 — Visual activism graphic

are things we do with those cultural forms to make changes, small or large, from a direct political action to a performance – whether in everyday life or in a theatre – a conversation or a work of art. Once we have learned how to see the world, we have taken only one of the required steps. The point is to change it.

Further Reading

ABBAS, ACKBAR (2012)
'Faking Globalization', in Mirzoeff (2012)

ABBATE, JANET (1999)
Inventing the Internet (Cambridge, MA: MIT Press)

ABEL, ELIZABETH (2010)
Signs of the Times: The Visual Politics of Jim Crow (Berkeley: University of California Press)

ADAMS, RACHEL (2001)
Sideshow USA (Chicago: University of Chicago Press)

AGGER, BEN (2012)
Oversharing: Presentations of Self in the Internet Age (New York: Routledge)

ANDERSON, BENEDICT (1991)
Imagined Communities: Reflections on the Origin and Spread of Nationalism (London: Verso)

AUDUBON, JOHN JAMES (1999)
Writings and Drawings (Washington, DC: Library of America)

AUDUBON SOCIETY (2007)
'Common Birds in Decline'
http://birds.audubon.org/common-birds-decline

AW, TASH (2013)
Five Star Billionaire (London: Fourth Estate)

AZOULAY, ARIELLA (2008)
The Civil Contract of Photography (New York: Zone Books)

BAVELIER LAB
'The Brain and Learning'
http://cms.unige.ch/fapse/people/bavelier

BEAUVOIR, SIMONE DE (1947)
The Second Sex (New York: Norton)

BENJAMIN, WALTER (1968)
'The Work of Art in the Age of Mechanical Reproduction', tr.
Harry Zohn (from a 1935 essay), in *Illuminations*, ed. Hannah
Arendt (New York: Schocken)

BENJAMIN, WALTER (1999)
The Arcades Project (Cambridge, MA: Belknap Press)

BERGER, JOHN (1973)
Ways of Seeing (Harmondsworth: Pelican)

BERGER, MAURICE (2010)
*For All the World to See: Visual Culture and the Struggle for Civil
Rights* (New Haven: Yale University Press)

BLUM, ANDREW (2013)
'Children of the Drone'
www.vanityfair.com/culture/2013/06/new-aesthetic-james-
bridle-drones

BOGGS, GRACE LEE (2011)
*The Next American Revolution: Sustainable Activism for the
Twenty-First Century* (Berkeley: University of California Press)

BRIDLE, JAMES (2012)
Dronestagram: The Drone's-Eye View
http://booktwo.org/notebook/dronestagram-drones-eye-view

BRYSON, NORMAN, MICHAEL ANN HOLLY AND KEITH MOXEY (EDS.)
(1994)
Visual Culture: Images and Interpretations (Hanover, NH:
Wesleyan University Press)

BUCK-MORSS, SUSAN (1992)
'Aesthetics and Anaesthetics: Walter Benjamin's Artwork Essay
Reconsidered', *October*, vol. 62 (Autumn)

BUTLER, JUDITH (1990)
Gender Trouble (New York: Routledge)

CARLYLE, THOMAS (1840)
On Heroes and Hero-Worship (London)

CARSON, RACHEL (1962)
Silent Spring (New York: Houghton Mifflin)

CASTELLS, MANUEL (1996)
The Rise of the Network Society (Oxford: John Wiley-Blackwell)

CERTEAU, MICHEL DE (1984)
The Practice of Everyday Life, tr. Steven Rendall (Berkeley: University of California Press)

CHABRIS, CHRISTOPHER, AND DANIEL SIMONS (2010)
The Invisible Gorilla: How Our Senses Deceive Us (New York: Harmony)

CHUN, WENDY HUI KYONG (2006)
Control and Freedom: Power and Paranoia in the Age of Fiber Optics (Cambridge, MA: MIT Press)

CHUN, WENDY HUI KYONG (2011)
Programmed Visions: Software and Memory (Cambridge, MA: MIT Press)

CLARK, ROY PETER (2013)
'Me, My Selfie and I'
http://edition.cnn.com/2013/11/23/opinion/clark-selfie-word-of-year

CLARK, T. J. (1973)
The Image of the People: Gustave Courbet and the 1848 Revolution (London: Thames & Hudson)

CLAUSEWITZ, CARL VON ([1832] 2006)
On War, tr. J. J. Graham (Project Gutenberg)

CNA MILITARY ADVISORY BOARD (2014)
National Security and the Accelerating Risks of Climate Change (Alexandria, VA: CNA Corporation)

COAL + ICE
http://sites.asiasociety.org/coalandice

COLE, ERNEST (1967)
House of Bondage (New York: Random House)

COLL, STEVE (2014)
'The Unblinking Stare'
http://www.newyorker.com/magazine/2014/11/24/
unblinking-stare

COLUMBIA LAW SCHOOL (2012)
Counting Drone Strike Deaths (New York: Human Rights Clinic,
Columbia Law School)

COMOLLI, JEAN-LOUIS (1980)
'Machines of the Visible', in Teresa de Lauretis and Stephen
Heath (eds.), *The Cinematic Apparatus* (London & Basingstoke:
Macmillan)

COULDRY, NICK, AND NATALIE FENTON (2011)
'Occupy: Rediscovering the General Will in Hard Times'
http://www.possible-futures.org/2011/12/22/rediscovering-
the-general-will

CRARY, JONATHAN (2014)
24/7 (New York: Zone)

DAVIS, MIKE (2006)
Planet of Slums (New York: Verso)

DELEUZE, GILLES (1992)
'Postscript on the Societies of Control', *October*, vol. 59 (Winter)

DER DERIAN, JAMES (2009)
*Virtuous War: Mapping the Military-Industrial-Media-Entertainment-
Network* (New York: Routledge)

DESCARTES, RENÉ (1637)
Discourse on Method (as *Discours de la méthode pour bien conduire
sa raison, et chercher la vérité dans les sciences*) (Paris)

DI JUSTO, PATRICK (2013)
'Object of Interest: Pepper Spray'
http://www.newyorker.com/tech/elements/object-of-interest-pepper-spray

EDWARDS, PAUL (1996)
The Closed World: Computers and the Politics of Discourse in Cold War America (Minneapolis: University of Minnesota Press)

EDWARDS, PAUL (2010)
A Vast Machine: Computer Models, Climate Data, and the Politics of Global Warming (Cambridge, MA: MIT Press)

EMARKETER (2013)
'Digital Set to Surpass TV in Time Spent with US Media'
http://www.emarketer.com/article/digital-set-surpass-tv-time-spent-with-us-media/1010096#sthash.tewzdbeq.dpuf

FANON, FRANTZ (1967)
Black Skin, White Masks, tr. Charles Lam Markmann (New York: Grove)

FELLEMAN, D. J., AND D. C. VAN ESSEN (1991)
'Distributed Hierarchical Processing in the Primate Cerebral Cortex', *Cerebral Cortex*, Jan/Feb, vol. 1, no. 1: pp. 1–47
http://www.ncbi.nlm.nih.gov/pubmed/1822724

FM 3-24 (2006)
Counterinsurgency (Washington, DC: US Army and Marine Corps)

FOUCAULT, MICHEL (1970)
The Order of Things, tr. A. M. Sheridan Smith (London: Tavistock)

FRIEDAN, BETTY (1963)
The Feminine Mystique (New York: Norton)

FRIEDBERG, ANNE (1994)
Window Shopping: Cinema and the Postmodern (Berkeley: University of California Press)

GALLESE, VITTORIO (2003)
'The manifold nature of interpersonal relations: the quest for a common mechanism', *Philosophical Transactions of the Royal Society B* (2003): pp. 358, 517–28

GALLOWAY, ALEXANDER R. (2012)
The Interface Effect (New York: Polity)

GANZEER (2011)
'Practical Advice', *Bidoun #25* (Summer): pp. 39–43

GANZEER (2014)
'Concept Pop', *The Cairo Review of Global Affairs*, 6 July 2014 http://www.aucegypt.edu/gapp/cairoreview/Pages/articleDetails.aspx?aid=618

GIBSON, WILLIAM (1984)
Neuromancer (New York: HarperCollins)

GOLDBLATT, DAVID ([1966] 1975)
Some Afrikaners Photographed (Johannesburg: Murray Crawford)

GOUREVITCH, PHILIP, AND ERROL MORRIS (2008)
Standard Operating Procedure DVD

GRAEBER, DAVID (2013)
The Democracy Project: A History, A Crisis, A Movement (New York: Random House)

GRUSIN, RICHARD (2010)
Premediation: Affect and Mediality After 9/11 (New York: Palgrave)

HALBERSTAM, JACK (2012)
Gaga Feminism: Sex, Gender and the End of the Normal (Boston: Beacon)

HALBERSTAM, JACK (2013)
'Charming for the Revolution: A Gaga Manifesto', *e-flux #44* (April 2013), http://www.e-flux.com/journal/charming-for-the-revolution-a-gaga-manifesto

HALBERSTAM, JUDITH (1998)
Female Masculinity (Durham, NC: Duke University Press)

HARVEY, DAVID (2013)
Rebel Cities: From the Right to the City to the Urban Revolution
(New York: Verso)

HOGAN, WESLEY C. (2009)
Many Minds, One Heart: SNCC's Dream for a New America (Chapel
Hill, NC: University of North Carolina Press)

JACQUES, MARTIN (2011)
When China Rules the World (New York: Penguin)

LACAN, JACQUES (2007)
'The Mirror Stage as Formative of the Function of the I as
Revealed in Psychoanalytic Experience', in *Écrits*, tr. Bruce Fink
(New York: Norton): pp. 75–82

LASCH, PEDRO (2010)
Black Mirror/Espejo Negro (Durham, NC: Nasher Museum of Art)

LLOYD, ANG (2014)
'Zanele Muholi's New Work Mourns and Celebrates South
African Queer Lives', *Africa Is a Country*, 20 March 2014
http://africasacountry.com/zanele-muholis-new-work-mourns-
and-celebrates-south-african-queer-lives

LOIPERDINGER, MARTIN, AND BERND ELZER (2004)
'Lumière's Arrival of the Train: Cinema's Founding Myth', *The
Moving Image*, vol. 4, no. 1 (Spring): pp. 89–118

LOSH, ELIZABETH (2014)
'Beyond Biometrics: Feminist Media Theory Looks at Selfiecity'
http://d25rsf93iwlmgu.cloudfront.net/downloads/liz_losh_
beyondbiometrics.pdf

LYOTARD, JEAN-FRANÇOIS (1979)
The Postmodern Condition (Minneapolis: University of Minnesota
Press)

MCLUHAN, MARSHALL (1962)
The Gutenberg Galaxy: The Making of Typographic Man (Toronto:
University of Toronto Press)

MCLUHAN, MARSHALL (1964)
Understanding Media: The Extensions of Man (New York: McGraw-Hill)

MCLUHAN, MARSHALL, AND Q. FIORE WITH J. AGEL (1967)
The Medium is the Massage: An Inventory of Effects (New York: Random House)

MARCHE, STEPHEN (2013)
'Sorry, Your Selfie Isn't Art'
http://www.esquire.com/blogs/culture/selfies-arent-art

MARX, KARL (1867)
'Notes for an Undelivered Speech on Ireland'
http://www.marxists.org/archive/marx/iwma/documents/1867/irish-speech-notes.htm

MATURANA, H. R. (1980)
'Biology of Cognition', in H. R. Maturana and F. J. Varela, *Autopoiesis and Cognition* (Dordrecht: D. Reidel): pp. 2–58

MAVOR, CAROL (1999)
Becoming: The Photographs of Clementina, Viscountess Hawarden (Durham, NC: Duke University Press)

MIÉVILLE, CHINA (2009)
The City and the City (New York: Del Rey)

MILLWARD, STEPHEN (2014)
'China's 450 million online video viewers watch 57 billion hours of vids every month' https://www.techinasia.com/china-has-450-million-online-video-viewers-2013-infographic

MIRZOEFF, NICHOLAS (2005)
Watching Babylon: The War in Iraq and Global Visual Culture (London: Routledge)

MIRZOEFF, NICHOLAS (2012)
The Visual Culture Reader, 3rd edition (London: Routledge)

MITCHELL, W. J. T. (2005)
'There Are No Visual Media', *Journal of Visual Culture*, August 2005, vol. 4, no. 2: pp. 257–66

MITCHELL, W. J. T. (2011)
Cloning Terror: The War of Images, 9/11 to the Present (Chicago: University of Chicago Press)

MULVEY, LAURA (1975)
'Visual Pleasure and Narrative Cinema', *Screen* 163 (Autumn): pp. 6–18

MUMFORD, LEWIS (1961)
The City in History: Its Origins, Its Transformations, and Its Prospects (Orlando, FL: Harcourt)

NASSI, JONATHAN J., AND EDWARD M. CALLAWAY
'Parallel Processing Strategies of the Primate Visual System', *Nature Reviews Neuroscience* 10 (1 May 2009): pp. 360–72

NIXON, ROB (2011)
Slow Violence and the Environmentalism of the Poor (Cambridge, MA: Harvard University Press)

NORA, PIERRE (2006)
Realms of Memory: Rethinking the French Past (New York: Columbia University Press)

ORESKES, NAOMI, AND ERIK CONWAY (2010)
Merchants of Doubt: How a Handful of Scientists Obscured the Truth on Issues from Tobacco Smoke to Global Warming (New York: Bloomsbury)

ORWELL, GEORGE (1937)
The Road to Wigan Pier (London: Gollancz)

PARKER, ROZSIKA, AND GRISELDA POLLOCK (1981)
Old Mistresses: Women, Art and Ideology (London: Routledge & Kegan Paul)

PARKS, LISA (2005)
Cultures in Orbit (Durham, NC: Duke University Press)

PHELPS, EARLE B., GEORGE A. SOPER AND RICHARD H. GOULD (1934)
'Studies of Pollution of New York Harbor and the Hudson River', *Sewage Works Journal* 6, no. 5: pp. 998–1008

PILLAY, DEVAN (2013)
'The Second Phase – Tragedy or Farce?', in *New South African Review 3* (Johannesburg: University of Witwatersrand Press)

RAMACHANDRAN, V. S. (2011)
The Tell-Tale Brain: A Neuroscientist's Quest for What Makes Us Human (New York: Norton)

RANCIÈRE, JACQUES (2001)
'Ten Theses on Politics', tr. Davide Panigia and Rachel Bowlby, *Theory & Event*, vol. 5, no. 3

RUDWICK, MARTIN J. S. (2005)
Bursting the Limits of Time: The Reconstruction of Geohistory in the Age of Revolution (Chicago: University of Chicago Press)

SCHECHNER, RICHARD (2002)
Performance Studies (New York: Routledge)

SCHIVELBUSCH, WOLFGANG (1987)
The Railway Journey: The Industrialization and Perception of Time and Space in the Nineteenth Century (Berkeley: University of California Press)

SCHMIDT, ERIC, AND JARED COHEN (2013)
The New Digital Age: Reshaping the Future of People, Nations and Business (New York: Knopf)

SELFIECITY
Investigating the style of self-portraits (selfies) in five cities across the world
http://selfiecity.net

SHIRKY, CLAY (2008)
Here Comes Everybody: The Power of Organizing Without Organizations (New York: Penguin)

SITRIN, MARINA (2006)
Horizontalism: Voices of Popular Power in Argentina (Oakland, CA: AK Press)

SNYDER, JOEL (1985)
'*Las Meninas* and the Mirror of the Prince', *Critical Inquiry*, vol. 11, no. 4 (June): pp. 539–73

SPERI, ALICE (2014)
'International and Egyptian Street Artists Join Forces Against Sisi', *VICE News*
https://news.vice.com/article/international-and-egyptian-street-artists-join-forces-against-sisi

STAROSIELSKI, NICOLE (2012)
'"Warning: Do Not Dig": Negotiating the Visibility of Critical Infrastructures', *Journal of Visual Culture* (April), vol. 11, no. 1: pp. 38–57

STIGLITZ, JOSEPH (2011)
'Of the 1%, by the 1%, for the 1%'
http://www.vanityfair.com/society/features/2011/05/top-one-percent-201105

SZE TSUNG LEONG (2004)
'History Images'
http://www.szetsungleong.com/texts_historyimages.htm

THOMPSON, E. P. (1991)
'Time, Work-Discipline and Industrial Capitalism', in E. P. Thompson, *Customs in Common: Studies in Popular Culture* (New York: The New Press)

VERTOV, DZIGA (1984)
Kino Eye: The Writings of Dziga Vertov (Berkeley: University of California Press)

VIRNO, PAOLO (2004)
A Grammar of the Multitude, tr. Isabella Bertoletti and James Cascaito (New York: Sémiotext(e))

VLADISLAVIC, IVAN (2009)
Portrait with Keys: The City of Johannesburg Unlocked (New York: Norton)

WALLACE, DAVID FOSTER (2007)
'Deciderization 2007 – a Special Report', Introduction to *Best American Essays 2007* (New York: Mariner Books)

WEIZMAN, EYAL (2007)
Hollow Land: Israel's Architecture of Occupation (London: Verso)

WERREL, CAITLIN E., AND FRANCESCO FEMIA (2013)
The Arab Spring and Climate Change, The Center for
Climate Security
https://climateandsecurity.files.wordpress.com/2012/04/
climatechangearabspring-ccs-cap-stimson.pdf

WUEBBLES, DONALD (2012)
'Celebrating Blue Marble', *Eos*, vol. 93, no. 49 (December):
pp. 509–10

Illustrations

Notes

CHAPTER 1: HOW TO SEE YOURSELF

1. Foucault [1966] 1970, pp. 14–15.
2. Ibid., pp. 16–17.
3. Lasch 2010, p. 10.
4. Parker and Pollock 1981, p. 99.
5. de Beauvoir 1947, p. 283.
6. Mulvey 1975, p. 33.
7. Schechner 2002, p. 29.
8. Butler 1990, p. xxii.
9. Fanon 1967, pp. 93–112.
10. Fosso, http://www.theguardian.com/artanddesign/2011/jun/19/photographer-samuel-fosso-best-shot
11. Halberstam 2013.
12. Clark 2013, http://edition.cnn.com/2013/11/23/opinion/clark-selfie-word-of-year
13. Marche 2013, http://www.esquire.com/blogs/culture/selfies-arent-art
14. Losh 2014, http://d25rsf93iwlmgu.cloudfront.net/downloads/liz_losh_beyondbiometrics.pdf

---

CHAPTER 2: HOW WE THINK ABOUT SEEING

1. Felleman and Van Essen 1991, http://www.ncbi.nlm.nih.gov/pubmed/1822724
2. Ibid., p. 30.
3. Ramachandran 2011, p. 55.
4. Ibid., p. 47.

5. Ibid., p. 124.
6. Ibid., p. 117.

CHAPTER 3: THE WORLD OF WAR

1. von Clausewitz [1832] 2006, p. 54.
2. Ibid., p. 38.
3. Ibid, p. 9. The translation given here is 'War is only a continuation of state policy by other means'.
4. Harlan K. Ullman and James P. Wade, *Shock and Awe: Achieving Rapid Dominance* (National Defense University, 1996), available at http://archive.org/stream/shockandaweachie07259gut/skawe10.txt
5. *Harper's* 2013, http://www.harpers.org/harpers-index/?s=drones
6. Coll 2014, http://www.newyorker.com/magazine/2014/11/24/unblinking-stare
7. Blum 2013, www.vanityfair.com/culture/2013/06/new-aesthetic-james-bridle-drones

CHAPTER 4: THE WORLD ON SCREEN

1. Vertov 1984, p. 40.
2. Ibid., p. 14.
3. McLuhan and Fiore 1967, p. 26.
4. eMarketer 2013, http://www.emarketer.com/article/digital-set-surpass-tv-time-spent-with-us-media/1010096#sthash.tewzdbeq.dpuf
5. Millward 2014, https://www.techinasia.com/china-has-450-million-online-video-viewers-2013-infographic
6. Starosielski 2012, pp. 38–57.
7. http://www.submarinecablemap.com/
8. Wallace 2007, p. 6.
9. http://www.eurofighter.com/news-and-events/2005/06/helmet
10. http://www.bbc.com/news/technology-19372299
11. Galloway 2012, pp. 42–8.
12. http://press.ihs.com/press-release/design-supply-chain-media/soaring-esports-viewership-driven-online-video-platforms
13. Deleuze 1992, p. 3.

14. http://www.google.com/glass. This picture has now been removed. See http://www.theguardian.com/technology/2013/apr/30/google-glass-pictures-online
15. Schmidt and Cohen 2013, p. 98.

CHAPTER 5: WORLD CITIES, CITY WORLDS

1. http://www.who.int/gho/urban_health/situation_trends/urban_population_growth_text/en/
2. http://www.chengduinvest.gov.cn/en/htm/detail.asp?id=12607
3. http://press.parisinfo.com/key-figures/key-figures/tourism-in-paris-key-figures-2013
4. Balzac, *The Lesser Bourgeois of Paris*, Ch. 1 (1855).
5. Baudelaire, *Flowers of Evil* (1857).
6. Baudelaire, *The Painter of Modern Life* (1863).
7. *Journal des Débats* (1831), quoted by Benjamin (1999), p. 35.
8. Hogan (2009), p. 35.
9. Pillay (2013), p. 12.
10. Nora (2006).
11. Sze Tsung Leong 2004, http://www.szetsungleong.com/texts_historyimages.htm
12. Aw (2013).
13. http://www.telegraph.co.uk/technology/10172298/one-surveillance-camera-for-every-11-people-in-britain-says-cctv-survey.html
14. Clement Valla, http://www.postcards-from-google-earth.com/info/

CHAPTER 6: THE CHANGING WORLD

1. Bacon ([1605] 2001), Book 1, Vol. 11.
2. Audubon 1999, pp. 263–4.
3. Audubon Society (2007), http://birdsaudubonorg/common-birds-decline 2007
4. Buck-Morss 1992, pp. 3–41.
5. Phelps et al. 1934, p. 1006.
6. Orwell 1937, p. 18.
7. http://www.worldbank.org/en/topic/poverty/overview

8. http://www.spiegel.de/international/world/chinese-leaders-forced-to-counter-environmental-pollution-a-886901.html

9. https://www.gov.uk/government/uploads/system/uploads/attachment_data/file/295968/20140327_2013_uk_greenhouse_gas_emissions_provisional_figures.pdf

10. https://www.gov.uk/government/uploads/system/uploads/attachment_data/file/261692/consumption_emissions_28_Nov_2013.pdf

11. http://www.bloomberg.com/news/2013-01-18/china-s-steel-production-rises-3-1-in-2012-as-economy-expanded.html and http://www.steel.org/about%20aisi/statistics.aspx

12. http://www.oecd.org/sti/ind/item%203.%20mckinsey%20-%20competitveness%20in%20the%20steel%20industry%20%28oecd%29%20-%20final.pdf

13. http://anishkapoor.com/332/orbit.html

14. http://www.prixpictet.com/portfolios/earth-shortlist/sammy-baloji/statement/

15. https://www.sec.gov/about/laws/wallstreetreform-cpa.pdf

16. http://sites.asiasociety.org/coalandice

CHAPTER 7: CHANGING THE WORLD

1. http://enlacezapatista.ezln.org.mx/sdsl-en

2. Sitrin 2006, pp. 3–5.

3. Werrel and Femia 2013, https://climateandsecurity.files.wordpress.com/2012/04/climatechangearabspring-ccs-cap-stimson.pdf

4. Couldry and Fenton 2011, http://www.possible-futures.org/2011/12/22/rediscovering-the-general-will

5. http://www.unicef.org/infobycountry/egypt_statistics.html

6. Ganzeer 2014, http://www.aucegypt.edu/gapp/cairoreview/Pages/articleDetails.aspx?aid=618

7. Ganzeer 2011, pp. 39–43.

8. http://www.ganzeer.com

9. http://muftah.org/freedom-of-expression-under-threat-in-north-africa-an-open-letter-from-ganzeer/#.vhxzttgsxto

10. For now, Egyptian graffiti can still be seen at http://suzeeinthecity.wordpress.com

11. Speri 2014, https://news.vice.com/article/international-and-egyptian-street-artists-join-forces-against-sisi

12. http://mosireen.org
13. youtu.be/47ipxanddtg
14. youtu.be/kvo3nqfkmbm
15. https://www.adbusters.org/blogs/adbusters-blog/occupywallstreet.html
16. Stiglitz 2011, http://www.vanityfair.com/society/features/2011/05/top-one-percent-201105
17. di Justo, http://www.newyorker.com/tech/elements/object-of-interest-pepper-spray
18. For South Africa: http://www.indexmundi.com/g/r.aspx?v=2229
    For Detroit: http://www.dennis-yu.com/calling-dan-gilbert-our-solution-to-detroits-youth-unemployment-issue/.
    For Palestine: http://www.pcbs.gov.ps/site/512/default.aspx?tabid=512&lang=en&itemid=790&mid=3172&wversion=staging
19. CNA Military Advisory Board 2014.

Index

Economics:
The User's Guide
Ha-Joon Chang

What is economics?

What can – and can't – it explain about the world?

Why does it matter?

Ha-Joon Chang teaches economics at Cambridge University and writes a column for the *Guardian*. The *Observer* called his book *23 Things They Don't Tell You About Capitalism*, which was a no.1 best-seller, 'a witty and timely debunking of some of the biggest myths surrounding the global economy'. He won the Wassily Leontief Prize for advancing the frontiers of economic thought and is a vocal critic of the failures of our current economic system.

A PELICAN
INTRODUCTION

Human Evolution
Robin Dunbar

What makes us human?

How did we develop language, thought and culture?

Why did we survive, and other human species fail?

Robin Dunbar is an evolutionary anthropologist and Director of the Institute of Cognitive and Evolutionary Anthropology at Oxford University. His acclaimed books include *How Many Friends Does One Person Need?* and *Grooming, Gossip and the Evolution of Language*, described by Malcolm Gladwell as 'a marvellous work of popular science'.

**A PELICAN
INTRODUCTION**

Revolutionary Russia, 1891–1991
Orlando Figes

What caused the Russian Revolution?

Did it succeed or fail?

Do we still live with its consequences?

Orlando Figes teaches history at Birkbeck, University of London and is the author of many acclaimed books on Russian history, including *A People's Tragedy*, which *The Times Literary Supplement* named as one of the '100 most influential books since the war', *Natasha's Dance*, *The Whisperers*, *Crimea* and *Just Send Me Word*. The *Financial Times* called him 'the greatest storyteller of modern Russian historians'.

A PELICAN
INTRODUCTION

The
Domesticated
Brain
Bruce Hood

Why do we care what others think?

What keeps us bound together?

How does the brain shape our behaviour?

Bruce Hood is an award-winning psychologist who has taught and researched at Cambridge and Harvard universities and is currently Director of the Cognitive Development Centre at the University of Bristol. He delivered the Royal Institution's Christmas Lectures in 2011 and is the author of *The Self Illusion* and *Supersense*, described by *New Scientist* as 'important, crystal clear and utterly engaging'.

A PELICAN
INTRODUCTION

Greek and
Roman
Political Ideas
Melissa Lane

**Where do
our ideas
about politics
come from?**

**What can we
learn from the
Greeks and
Romans?**

**How should
we exercise
power?**

Melissa Lane teaches politics
at Princeton University, and
previously taught for fifteen
years at Cambridge University,
where she also studied as a
Marshall and Truman scholar.
The historian Richard Tuck
called her book *Eco-Republic*
'a virtuoso performance by one
of our best scholars of ancient
philosophy'.

A PELICAN
INTRODUCTION

Classical
Literature
Richard Jenkyns

**What makes
Greek and Roman
literature great?**

**How has
classical literature
influenced
Western culture?**

**What did Greek
and Roman
authors learn
from each other?**

Richard Jenkyns is emeritus
Professor of the Classical
Tradition and the Public
Orator at the University of
Oxford. His books include
Virgil's Experience and *The
Victorians and Ancient Greece*,
acclaimed as 'masterly' by
History Today.

A PELICAN
INTRODUCTION

Who Governs Britain?
Anthony King

Where does power lie in Britain today?

Why has British politics changed so dramatically in recent decades?

Is our system of government still fit for purpose?

Anthony King is Millennium Professor of British Government at the University of Essex. A Canadian by birth, he broadcasts frequently on politics and government and is the author of many books on American as well as British politics. He is co-author of the bestselling *The Blunders of Our Governments*, which David Dimbleby described as 'enthralling' and Andrew Marr called 'an astonishing achievement'.

A PELICAN
INTRODUCTION